HBR's 10 Must Reads

UPDATED &
EXPANDED

I0112626

Leading Digital Transformation

HBR's 10 Must Reads

HBR's 10 Must Reads are definitive collections of classic ideas, practical advice, and essential thinking from the pages of *Harvard Business Review*. Exploring topics like disruptive innovation, emotional intelligence, and new technology in our ever-evolving world, these books empower any leader to make bold decisions and inspire others.

TITLES INCLUDE:

HBR's 10 Must Reads for New Managers
HBR's 10 Must Reads on Artificial Intelligence
HBR's 10 Must Reads on Building a Great Culture
HBR's 10 Must Reads on Change Management
HBR's 10 Must Reads on Communication
HBR's 10 Must Reads on Data Strategy
HBR's 10 Must Reads on Decision-Making
HBR's 10 Must Reads on Emotional Intelligence
HBR's 10 Must Reads on High Performance
HBR's 10 Must Reads on Innovation
HBR's 10 Must Reads on Leadership
HBR's 10 Must Reads on Leading Digital Transformation
HBR's 10 Must Reads on Leading Winning Teams
HBR's 10 Must Reads on Managing People
HBR's 10 Must Reads on Managing Yourself
HBR's 10 Must Reads on Marketing
HBR's 10 Must Reads on Mental Toughness
HBR's 10 Must Reads on Strategy
HBR's 10 Must Reads on Women and Leadership
HBR's 10 Must Reads Boxed Set (6 books)
HBR's 10 Must Reads Ultimate Boxed Set (14 books)

For a full list, visit hbr.org/mustreads.

HBR's 10 Must Reads

UPDATED & EXPANDED

Leading Digital Transformation

Harvard Business Review Press
Boston, Massachusetts

Copyright 2026 Harvard Business Publishing Corporation

All rights reserved

Printed in the United States of America

10 9 8 7 6 5 4 3 2 1

No part of this publication may be reproduced, stored in or introduced into a retrieval system, or transmitted, in any form, or by any means (electronic, mechanical, photocopying, recording, or otherwise), without the prior permission of the publisher. Requests for permission should be directed to permissions@harvardbusiness.org, or mailed to Permissions, Harvard Business School Publishing, 60 Harvard Way, Boston, Massachusetts 02163.

The web addresses referenced in this book were live and correct at the time of the book's publication but may be subject to change.

Cataloging-in-Publication data is forthcoming.

ISBN: 979-8-89279-292-9
eISBN: 979-8-89279-293-6

The paper used in this publication meets the requirements of the American National Standard for Permanence of Paper for Publications and Documents in Libraries and Archives Z39.48-1992.

Contents

7

A Better Way to Put Your Data to Work   125
Package it the way you would a product.

**by Veeral Desai, Tim Fountaine, and
Kayvaun Rowshankish**

8

The Age of Continuous Connection   149
Build deeper, more customized ties with your
customers, or your competitors will.

by Nicolaj Siggelkow and Christian Terwiesch

9

Want Your Company to Get Better at Experimentation?   163
Empower everyone to test and refine new ideas.

**by Iavor Bojinov, David Holtz, Ramesh Johari,
Sven Schmit, and Martin Tingley**

10

Reskilling in the Age of AI   175
Insights from nearly 40 organizations with
large-scale reskilling programs.

**by Jorge Tamayo, Leila Doumi, Sagar Goel,
Orsolya Kovács-Ondrejkovic, and Raffaella Sadun**

Leading Digital Transformation

1

Discovery-Driven Digital Transformation

by Rita McGrath and Ryan McManus

What's your digital strategy? That simple question often throws the CEOs of traditional companies into a panic. They believe that digital technologies and business models pose an existential threat to their way of doing business—and of course they're right. But the pressure they feel often leads them to make big bet-the-farm moves—and that's usually wrong.

Veon, a large multinational provider of telecommunications services, is a case in point. Its new digital platform, introduced in 2017, was a huge project, involving 100 staff members in Amsterdam and another hundred or so in its London office. The idea was to create a mobile app that would offer users rich localized experiences and serve as a sales channel for Veon's commercial partners (such as Mastercard). Management considered the project its top priority. But after being launched with much fanfare, the app got a lukewarm response from customers, and the effort to build a new ecosystem around it was scrapped. The failure led to a management exodus, layoffs, and a back-to-basics strategy with digital efforts sidelined to pilot-project stage.

Veon still needs a new business model, though, and clearly can't afford to make many more large investments in searching for one.

It doesn't have to. Just because a threat is huge doesn't mean that a response has to be. To the contrary, companies like Veon would actually be much better off taking a more incremental approach to transformation over time. While they should always have a vision of where they want to go, they should work their way toward it by continually finding opportunities to digitize problematic processes in their core operations. When they tackle those projects, they'll learn what metrics to use, which assumptions to revise, where they can introduce new business models, and who their new competitors might be. And as they absorb those lessons, their understanding of their competitive landscape—and the long-term goals they set for themselves—will inevitably change.

There's already a process for this kind of ongoing learning approach to strategy: discovery-driven planning (DDP). One of us, Rita, and Ian MacMillan developed it in the 1990s as a product innovation methodology, and it was later incorporated into the popular "lean startup" tool kit for launching businesses in an environment of high uncertainty. At its center is a low-cost process for quickly testing assumptions about what works, obtaining new information, and minimizing risks.

In the following pages we'll describe how an adapted form of DDP can help incumbent firms confront digital challenges and learn their way toward a new business model. Let's begin by looking in more detail at why a step-by-step transformation

Idea in Brief

The Problem

Established companies spend billions trying to turn themselves into digitized orchestrators of some new ecosystem, only to fall flat on their faces.

Why It Happens

The CEOs believe that the existential threat posed by digital disrupters requires a gigantic, model-busting response.

The Solution

Adopt an incremental experimental approach: discovery-driven digital transformation. Look for problems to fix with digital technology, but exploit your rich knowledge of customers, broad operational scope, and deep talent pools while learning your way to a new business model.

works better for traditional firms than the all-or-nothing approach that characterizes a startup's pivot.

The Incremental Advantage of Incumbent Firms

Economists have long puzzled over why firms exist at all and, at a more granular level, which tasks belong within the boundaries of a given firm. One line of thought, begun by Ronald Coase in the 1930s, suggests that under certain conditions, market transactions often are not satisfactory for individuals: when it is difficult or expensive to get information about what you want to buy, when bargains are hard to strike because information is asymmetrical, and when it's costly or challenging to enforce agreements. If any of those conditions apply, it makes sense to keep the activities involved within a firm.

Until fairly recently, the boundaries between firms and markets were well understood and relatively fixed. But digital technologies have changed all that by making it possible to use markets for a lot of work that once was done more efficiently within firms. Platforms such as Alibaba and Amazon have made it easy to outsource functions like selecting suppliers, negotiating prices, enforcing contracts, managing payments, and more.

As a result, executives in companies that were born digital have assumptions about how transactions should be structured that are completely different from those of executives in legacy companies. What's more, because digital firms' structures are evolving all the time, their managers revisit those assumptions frequently. Direct-to-consumer businesses (think Casper in mattresses, Harry's in shaving, and Warby Parker in eyeglasses) are constantly experimenting with and adjusting features like free shipping, product bundles, bonuses for adding items, and so on. Those tactics simply aren't available to an incumbent selling through distributors. And because the digital businesses cut out intermediaries, they can be profitable at a much lower scale.

A key consequence of all this is that digital startups can change direction, or pivot, without destroying much value. They usually aren't that capital-intensive and don't have big payrolls. The founders of Rooted, for instance, initially sold plants out of their apartment directly to consumers, only later moving to a separate space and hiring employees. For such companies, failure is relatively cheap—unless it happens late in the day (or investors succumb to the growth-at-all-costs mantra that is unraveling the fortunes of many so-called unicorns).

The employees, managers, and shareholders of traditional companies, however, cannot pivot without destroying value. If their digital gambles fail, workers lose their jobs, and physical assets have to be unloaded at fire-sale prices. And unlike the venture capitalists who back startups, the investors in what was once a safe company may not have the buffer of high-return investments to offset their losses.

But although incumbent firms can't pivot easily, the good news is that they don't need to. Think about what big companies can do that startups can't. Entrepreneurial ventures nearly always exploit a single idea. They usually can't try out multiple versions of the same idea at the same time, let alone multiple ideas. A big firm, in contrast, has the resources to explore a variety of ideas and can more easily experiment with different processes and operations, which makes it more likely to discover a dominant model than a startup is. This also gives a large firm a better chance of responding effectively to a digital challenge.

Take the case of the German metals distributor Klöckner. Its CEO, Gisbert Rühl, wanted to build a digital platform for the entire industry—but he didn't sponsor a big-bang effort to create one. Instead, his goal was to build digital competencies gradually, while benefiting from the knowledge and insight of people working in the firm's core steel business. For the first two years Rühl focused on digitizing inefficient manual processes; the firm created an online shop, a contract portal, order transparency tools, and a parts-manager app. Through these efforts it learned enough to create a platform on which the company and customers could seamlessly interact.

Klöckner's story reveals another advantage that incumbent firms have, at least in the early stages of an industry's adoption of digital models. They're led by people who already know their customers and can mine rich databases of prior transactions for insights. Startups are often led by technical experts and tend to be driven by new technical functionality rather than by the full portfolio of what customers are looking for. If you put a team of people who know the customers on the job, you'll stand a better chance of making your digital investment pay off. That's why Klöckner insisted that every project focus on how to help customers communicate more easily and efficiently with the company. That isn't the only goal to set, of course. Another company might start with a priority on shortening the time it takes to respond to a customer request. But whatever the goal is, it should frame the technology as an opportunity for the business rather than frame the business as an opportunity for the technology.

Once you accept the idea that firms should aim to disrupt in a nondisruptive manner, the challenge is subtly transformed from "What new business model should we back?" to the more nuanced question, "How can we learn our way toward a model that's right for our business?" That is where DDP comes in.

The Digital Context

DDP is somewhat like reverse engineering. When you use it in product development, you begin by imagining the offering you want to create and then figure out what you would need to change in order to get there. When you apply it to digital transformations, however, the focus is on reinventing the way you sell and deliver the products you already make as well as on identifying how to create and deliver new value through new digital capabilities.

Take power generation. Digital technologies are disrupting this once-stable industry, just as they are many other industries. Traditionally, power was generated from a central source and sent to its destination over a centrally managed grid. But new advances have made it possible to dynamically distribute power generated from dispersed small-scale producers tapping multiple energy sources. People with solar panels on their roofs or windmills in their gardens can sell surplus energy back to the grid, making households' cost of investing in power generation hardware more affordable and reducing the public's reliance on huge fossil-fuel power plants. If incumbents assume that the old business model will predict future success, they're likely to make big mistakes. General Electric's failed bet on the continued dominance of fossil-fuel-based electric plants provides a dramatic example.

Let's explore what's involved in applying a DDP approach to digital transformations. There are five key steps:

1. Define the operating experience: It's not just about digital

Before investing in a line of code, look for what isn't quite working in your operation. Where do you regularly need workarounds or have to stop a process unexpectedly to fetch more information or involve another person? These are likely to be areas that digitization can improve. Then think about how to redesign your operations there so that technology adds value, by making offerings and processes better, faster, cheaper, or more convenient.

The retailer Best Buy is one incumbent that was able to reconfigure its business operation in a way that created competitive advantages the digital-only players couldn't replicate. Back in 2010, Amazon released its price-comparison app, one of many tools that allowed shoppers to check out products in a physical store but order the same items at a discount online. Called "showrooming," the practice threatened to squeeze the lifeblood out of retail chains, which struggled to offer competitive prices while paying for real estate, staff, and inventory. It was one of the reasons Best Buy lost $1.7 billion in a single quarter in 2012.

Hubert Joly, the CEO hired to turn the company around, centered his strategy (and his business model) on solving two problems: negative comparable sales and declining operating margins. To do this, he envisioned a company that blended the human, the physical, and the digital in ways that an online-only player would find hard to match. He began by imagining what kind of customer experience Best Buy could deliver and, more important, identifying where it hadn't leveraged digital technologies to create that experience.

From this was born Best Buy's Renew Blue project, which had five components: a reinvigorated customer experience; a change in vendor partnerships; investments in ecological and social initiatives; the employee experience; and a return for investors. Financial targets and experiments were set up for each component.

To improve the employee experience, Best Buy launched initiatives focused on workforce morale, such as bringing back a popular employee discount that had been discontinued and investing in more-intensive training. To appeal to customers, the company began to match the prices of Amazon and other e-commerce players, which required a massive effort to overhaul Best Buy's warehousing, software, and supply chain activities. But because customers could walk out of the stores with the products, they could avoid the wait and the hassles (such as porch piracy) of having expensive products delivered, and that gave Best Buy an important edge. The company also created a system through which customers could order goods online for delivery or for pickup at the store. With 70% of Americans living within 15 miles of a Best Buy outlet, that approach proved to be extremely cost-effective.

Best Buy's new model turned the disadvantage of costly real estate into an advantage. At its more than 1,000 big-box locations, brands such as Apple, Samsung, and Microsoft created stores-within-stores, essentially paying rent to feature their offerings where real live shoppers could discover them in person. Best Buy is a neutral party to warring tech giants; archrivals Amazon and Google both sell their goods there. Finally, Best Buy invested in an in-home adviser capability, in which salaried, highly trained consultants go to customers' homes and provide tech help without selling anything. The goal is just to build stronger relationships with consumers. Throughout it all, Best Buy steadily transformed its digital footprint to support the strategy.

The Best Buy story illustrates the importance of being willing to rethink assumptions about how to use assets and engage with partners. Previous leaders in the firm had failed to see any way that it could price-match online retailers. But because Joly challenged traditional thinking, he spurred the company to re-imagine relationships with vendors (which now pay to be in Best Buy stores) and redesign its supply chain so that the company's physical assets could support a new business model for competing with e-commerce giants.

2. Focus on specific problems: Identify outcomes and progress metrics

The key question in any digital-transformation strategy is, How can we use data and digital capabilities to create new value for our customers? The DDP process translates that challenge into clear project goals.

A traditional success metric for new projects, even today, is return on investment. But ROI doesn't help you understand what value a project adds for customers, at least not directly. Further, to calculate it you need to estimate both investments and returns, which is precisely what you haven't figured out yet. What you need to do instead is identify metrics that are more closely linked to the specific improvements you hope digital initiatives will bring about.

We typically collect all this information in a "from-to" table, which identifies a problem, describes what a solution would achieve, and proposes a way to measure progress on that solution. (For an example, see the exhibit "Tackling big change step-by-step.") As you work through solving these problems, you'll test and refine your assumptions—a key DDP discipline. You can also capture what you've spent to gain new insights and what they've saved you. Eventually, you can back into something similar to an ROI calculation.

At Klöckner, the ultimate goal was to change the business model in steel from marking up inventory to a services revenue model. At first, the digital initiatives were simple and were focused on improving the order process—by, for instance, replacing the faxing of orders with an online portal for ordering. With each one, performance on metrics such as turnaround time and the number of steps required to complete an order improved. As the company gained more knowledge and capabilities, its projects became more ambitious.

Of course, you still need a way to measure progress on digital transformation overall, and to do that we suggest a metric we call *return on time invested* (ROTI). To calculate it, you simply divide your total revenue by the number of employees. The idea is that successful technology investments should let you accomplish more with fewer people. For example, we used annual report data from 2018 to compare Amazon (a digital-first company) with Walmart (a more-traditional legacy business). We found that Amazon had $232.9 billion in net sales and 647,500 full- and part-time workers. Its sales per employee were $359,671.

Tackling big change step-by-step

A key part of discovery-driven transformation is identifying organizational problems that can be addressed with digital technology, the desired improvement for each, and a metric for assessing progress toward it. All this information is captured in a tool called the from-to table. Below is how one financial services organization we've worked with filled out its table.

From	To	Progress metric
No consistent information about investments in a portfolio of projects; manual process	Clear and easily obtained information about investment flows; automated process	Reduction in time needed to update portfolio review information from 10 days to seconds
Significant effort needed to onboard new team members and bring them up to speed	Automated onboarding assistance that helps new team members learn the background of a project	Reduction in time it takes new team members to reach productivity from 30 days to five; high engagement scores among 85% of team members and top-quadrant scores for psychological safety
No capture of learning created in one part of the organization for reuse elsewhere	Routine recording of project insights in a database that's searchable by keywords, geographies, and contexts	Information is shared by an average of 10 units in the organization; 50% decrease in duplicative experiments
Slow and inflexible financial and talent resource allocation across new opportunities	Dynamic prioritization and resource allocation driven by real-time data and discovery	Resource reallocation cycles go from quarterly or annually to weekly, annual 50% to 100% increase in number of experiments with strategic options

In contrast, Walmart had $495.8 billion in net sales and 2.3 million associates. Its sales per employee were $215,548. Amazon enjoyed 67% higher performance per employee.

3. Identify your competition: Cast a wide net

Industry boundaries have blurred so much that standard industrial classification (SIC) codes are more or less useless. This by itself is one reason why conventional strategy-making approaches predicated on boundary assumptions are failing incumbents.

We suggest that leaders instead think about the field of competition not as a marketplace where similar players offer rival products and services but as what strategists call an arena. An arena is defined by a customer need—what Clay Christensen dubbed the "job to be done." It's a notion that goes back to Ted Levitt, who recommended that railway companies see themselves as competing in the transportation business against airlines, buses, trucking, and even cars. If railway passengers are a market, transportation users constitute an arena.

Smart born-digital firms already think this way. For example, Netflix has been very clear that it doesn't intend to compete just against television or the movies for viewers' time. It intends to compete against every possible leisure activity that a person might do instead of watching streaming content. The company sees traditional media companies as its rivals, of course, but its leadership looks at magazines, books, podcasts, and sporting events as competition as well.

At this point in the process, you should go back and determine whether the outcomes and metrics of success you spelled out in steps one and two are reasonable, given the arena you're competing in. Is your category losing share of wallet to others in the arena, for instance, or holding its own? Netflix has plenty of room to meet its growth goals, because total hours of video viewing are increasing and a lot of that growth is from streaming video.

4. Look for platforms: Don't forget the ecosystem implications

In the digital economy, striving to become an intermediary through which others buy and sell goods is an extremely popular strategy. It's a tempting business model, because once the two sides of a market have joined a platform they have little incentive to jump to another. This is partly due to network effects, whereby a platform's value to any user increases as the number of other users on that platform rises. Airbnb, for instance, benefits when more hosts and more guests use it and has historically gone to great lengths to ensure the loyalty of both.

A platform is also attractive because it needs less capital. To run a conventional hotel, you have to have real estate, rooms that need to be looked after, reservation systems, staff, and so on. Airbnb, in contrast, taps an ecosystem of hosts to provide all those things, and its directly controlled activities are simply to match hosts and guests and guarantee transactions, both of which occur entirely in the cloud and thus are infinitely scalable.

To understand whether a platform opportunity exists, we use a tool we call a *customer consumption chain* (introduced in HBR in 1997). The idea is simple: that as customers try to get jobs done in their lives, they go through a series of experiences, beginning with awareness of a need, then working through how to get that need met, and going all the way to the conclusion of a service or the end of a product's life. Digital technologies make open-market transactions for many links in that chain possible, allowing firms to build platforms.

That sounds like bad news for established organizations. But they have an ace in the hole: They employ many people who have deep technical expertise or understand customer problems. Those capabilities can give them an edge in identifying platformlike opportunities and building ecosystems. At Klöckner, Rühl realized that once there was price transparency—and far less friction—in the trading of basic metal commodities, competitive advantage would shift to suppliers that could offer superior solutions and service. The company blended the new ways of operating on platforms (co-creating designs with customers, for instance) of its digital arm with its workforce's deep expertise (in, say, manufacturing with 3-D lasers) to develop customized, higher-value offerings.

Becoming a popular platform isn't easy for corporations. The business landscape is littered with would-be platforms that failed even though they seemed to have all the right components. General Electric's Predix initiative, which was intended to be the platform for the industrial internet of things, is an example. Rather than driving the digitization of services that customers would value, Predix was sucked into serving primarily internal GE units—and a lot of them. Further, as part of GE Digital, the initiative was given P&L responsibility, which oriented it toward short-term contracts with customers that could pay some bills in the interim. It also took on way too much too soon, rather than proceeding by finding a good fit for its capabilities and building from there.

5. Test your assumptions: Failures are lessons too

One of the more popular tools to come out of DDP is the assumption checkpoint table. To create one, just write down the next few milestones that your digital project will go through, which assumptions need to be tested at each, and if possible, how much that test will cost. The beauty of this approach is that it moves the conversation from "Oh, you were wrong, that was a failure" to "Was it worth that price to learn what we needed to learn?"

Consider how Buffer, a service that allows people to space out social media promotions without having to predetermine the timing, tested assumptions in its launch phase. Joel Gascoigne, Buffer's cofounder, got the idea for the business from his own frustration with how clunky it was to try to tweet more consistently.

The first assumption he wanted to test was whether anybody else perceived this to be a problem. So he built a very simple two-page website. The first page's pitch was "Tweet More Consistently with Buffer." If users clicked on it, they were taken to a second page, with the heading "Hello, You Caught Us Before We're Ready," which had a place for people to enter their email addresses if they were interested in Buffer's solution. Most people weren't, but some were. So Gascoigne added a third page between the other two to test pricing hypotheses. And again, most people weren't interested in paying, but enough were to persuade Gascoigne to build the product.

Next he had to decide how complex to make it and how many social platforms to apply it to. He ended up keeping it very simple and supporting only Twitter at first. As of 2018, Buffer had more than 1.4 million social accounts connected to its apps.

Many large corporations have adopted a similar test-and-learn mindset. Several new services make experimentation easier—for example, Alpha, whose subscribers use it to obtain fast feedback about products from potential customers before making expensive or irreversible decisions. At WellMatch, an Aetna business unit, experimentation helped resolve disagreements about design decisions. According to Etugo Nwokah, the former chief product officer, one area of disagreement involved its website: Every group in the unit wanted to have its content appear on the landing page.

The trial entry page ended up being so busy that it confused consumers. The company had to go back to the drawing board and do a redesign—but was able to do so at a much lower cost and risk than if the webpage had been launched for real.

Conclusion

Digital transformation is complex and requires new ways of approaching strategy. Starting big, spending a lot, and assuming you have all the information is likely to produce a full-on attack from corporate antibodies—everything from risk aversion and resentment of your project to simple resistance to change.

A discovery-driven approach gets leaders past the common barriers to digital transformation. By starting small, spending a little on an ongoing portfolio of experiments, and learning a lot, you can win early supporters and early adopters. By then moving quickly and demonstrating clear impact on financial performance indicators, you can build a case for and learn your way into a digital strategy. You can also use your digitization projects to begin an organizational transformation. As people become more comfortable with the horizontal communications and activities that digital technologies enable, they will also embrace new ways of working.

Incumbent companies have some great advantages over new competitors: paying customers, financial resources, customer and market data, and larger talent pools. But CEOs will have to integrate agility and innovation into their broader organizations and communicate the new ways of digital thinking while minimizing disruption to their existing businesses. A discovery-driven approach provides a way to address those challenges.

Originally published in May–June 2020. Reprint R2003J

Is Your Company Squandering Digital Opportunities?

by Mohan Subramaniam

More than 60 years ago, Harvard Business School professor Theodore Levitt famously argued that companies often fail because they focus so narrowly on products and services that they forget to keep in mind the bigger picture: what consumers actually want. Levitt called this problem "marketing myopia," and it remains a problem to this day. Increasingly, however, companies are struggling with a new affliction, which I call *digital myopia*.

Digitally myopic firms insist on looking to products, services, and industry attributes for competitive advantage. They fail to notice that in today's world, customer preferences have shifted from these attributes to new data-driven services and experiences, and they fail to appreciate the growing value of data and the ways in which digital ecosystems can help them harness that data.

For the past few years, I've been extensively researching the topic of digital disruption, and in that work, I've identified the five main traps that firms need to watch out for if they hope to avoid digital myopia. In this article, I'll discuss those traps and suggest ways to overcome them.

The Product Trap

Firms caught in the product trap believe products to be their only revenue source, and they don't see the new and enlarged role that modern data can now occupy in their businesses. They rely only on *episodic data*—that is, data generated by discrete events, such as the shipment of a component or the sale of a product. Episodic data allows firms to monitor inventory levels or sales performance in different regions. That's important, but increasingly, firms today have an opportunity to gather and take advantage of *interactive data*—that is, data streamed continuously back to them as users interact with their products. Data was once used to support products, but now products can be used to support data.

Consider sensor-equipped smart inhalers, which demonstrate that shift. Smart inhalers can remind users to bring them on trips. They can prompt users to take their regularly prescribed doses. They can detect specific irritants, such as dust, pollen, or mold. For consumers, these data-driven features add convenience and value, and they can even save lives. For companies, they add new revenue streams.

To avoid product traps, firms need to start thinking about products as conduits for interactive data. Many firms are already doing this. Some embed microchips in their products (like smart inhalers); others use apps or websites. Allstate Insurance, for example, offers apps to track driving behavior, assess risks, and incentivize safety. Abilify, a medication for bipolar disorder, embeds ingestible sensors that enable relatives to ascertain dosage

Idea in Brief

The Problem

Some companies want to embrace digital transformation, but insist on looking to products, services, and industry attributes for competitive advantage. They fail to notice that customer preferences have shifted to new data-driven services and experiences and fail to appreciate the growing value of data and how digital ecosystems can help them harness it.

The Solution

Executives should avoid five traps that can limit the company's digital possibilities. The product trap involves a product-first rather than data-first mindset. The value-chain trap limits a firm's value chain to sales and after-sales servicing. The operational-efficiency trip includes thinking of digital gains only in terms of efficiency. The customer trap is the tendency to think of customers merely as people who buy products, rather than sources of interactive data. And in the competitor trap, a firm sees its rivals as companies that make similar products, not similar data.

The Payoff

When companies understand these traps and take steps to avoid them, they will be better positioned to seek out and capitalize on the full value that digital transformation offers.

compliance. Not all products can or should be equipped with sensors, of course, but firms must be open to the idea that what they offer consumers can change with new developments in the rapidly evolving world of sensors and the internet of things.

The Value-Chain Trap

Firms fall into the value-chain trap by believing that their value chains limit their business scope. Traditionally, firms have assumed that sales and after-sales servicing represent the end of their value chain.

But they don't. Consumers who've bought cars, after all, need roads, gas stations, and independent service providers. Consumers who've bought light bulbs need sockets, wiring, and electricity. In the past, legacy business models rarely took such product complements into account, because doing so didn't make business sense. But sensors and the internet of things have created opportunities for firms to expand their scope by doing just that.

How? By creating whole new *consumption ecosystems*—that is, networks that generate and share data and use it to connect product users to third-party entities that can offer product users additional related services. If your car has a sensor that monitors where you are and how full your tank is, when the time is right it can alert you that gas is running low and guide you to a nearby gas station. Streetlight bulbs with noise sensors can detect the sound of gunshots, turn on camera feeds, make 911 calls, and summon ambulances.

To participate in consumption ecosystems, firms must extend their value chains into digital platforms, which facilitate exchanges using real-time interactive data. Smart inhalers can monitor environmental triggers; toothbrushes can connect users to dentists and health insurers; vacuum cleaners can sense mouse droppings or termite activity and connect users to pest services.

To avoid falling into the value-chain trap, firms must develop processes to track new consumption ecosystems and find ways to build new digital platforms. Consider Dubai Ports, a shipping company whose traditional value-chain scope consisted of shipping goods through its port-to-port container services. But the firm is now planning to track emergent consumption ecosystems that include thousands of third-party entities that unload goods and make last-mile deliveries. It is also developing new digital platforms that can share real-time data, such as expected arrival times, with these third-party entities. This allows it to coordinate complementary activities after the goods land and expand the traditional value chain and its scope.

The Operational-Efficiency Trap

Consumption ecosystems may be unfamiliar to many firms, but *production* ecosystems are not.

In production ecosystems, firms turn their internal value-chain assets, processes, and entities into networks for generating and sharing data. They might simply use IT to automate order intakes or billing. Or they might go beyond that and use sensors, the internet of things, and artificial intelligence to create "lights-out factories" in which machines intelligently interact with one another and enable plants to run for weeks at a time with little human intervention—an innovation that can save millions of dollars. Firms might also generate data-driven services. That's what Caterpillar is doing: It has developed a variety of sensors and technologies to track wear-and-tear data on thousands of pieces of its equipment as they operate at hundreds of construction sites. This allows the firm to anticipate component failures and offer predictive maintenance, both of which help avoid costly delays.

Boosting operational efficiencies has its advantages, but if firms believe that operational-efficiency enhancements are the only—or the best—use of modern digital technologies, they may fall into the operational-efficiency trap. And if that happens, they'll underutilize their production ecosystems.

Avoiding the operational-efficiency trap requires that firms fully exploit the power of modern production ecosystems and design their business models accordingly. They have to figure out how to make their value chains into data-generating and data-sharing networks in ways that drive new services. The mattress producer Sleep Number has done that by creating smart mattresses that gather data on customers' heart rates and breathing patterns, which it then uses to track their sleep quality. The firm is now working on using its data to identify chronic sleep issues such as sleep apnea or restless-leg syndrome, which can predict heart attacks or strokes. Such data-driven services have made Sleep Number not just a mattress producer but also a wellness provider.

The Customer Trap

Firms fall into the customer trap when they think of customers only as people or groups who buy their products. Most legacy firms fall into this category: They have yet to recognize customers as sources of interactive data; they don't offer smart products; and they don't have plans to transform their legacy customers into digital customers.

Some firms fall into this trap because the status quo seems fine to them; they believe their product revenues are sufficient. Others believe that their economies of scale in manufacturing, branding, and distribution will maintain their competitive advantage. These sorts of beliefs can blind firms to the new opportunities and threats that data and digital ecosystems can introduce. They can also prevent firms from recognizing the rising power of network effects, which enhance the value of a product for a single user when that product is also used by many others. Again, consider smart inhalers: The larger the pool of digital customers and third-party data providers, the stronger the algorithms that the smart-inhaler firm can develop—and, in turn, the more precise the information and warnings that the firm can then send back to consumers.

To avoid the customer trap and reap the benefits that data and digital ecosystems have to offer, you first have to amass lots of data. That's not easy. Marshaling network effects thus becomes a priority, and doing so involves finding ways to incentivize and attract digital customers. Digital platforms such as Facebook and Google give their core platform away for free but generate revenues from select platform users, notably advertisers. Legacy firms must get creative and devise similar approaches, tailored to their business conditions. Before you declare this impossible in your industry, think about this: Just a decade ago, would you have imagined that firms in the business of making inhalers, mattresses, and farm equipment would all now have business models that involve monetizing the network effects created by their digital customers?

The Competitor Trap

Firms fall into the competitor trap when they believe their competitors to be only those that offer similar products and don't notice new digital competitors or competitors that offer similar data. That's what happened to legacy Chinese banks, which in recent years have allowed Alibaba and Tencent to usurp a sizable share of the market in loans with the help of their powerful digital platforms for e-commerce, search, payment services, and social networking. Although the legacy banks knew how to sell money for specific needs, Alibaba and Tencent used their digital platforms to understand the broader contexts in which people were using their money, and that gave them an advantage that the legacy banks didn't notice until it was too late.

To avoid falling into the competitor trap, firms must find ways to track their digital competitors. Not all will be digital platforms. Some may be startups, and others may be familiar product and industry rivals that have transformed themselves into digital competitors, as Oral-B (P&G) and Sonicare (Phillips) have done in the electric-toothbrush industry by producing smart toothbrushes that offer data-driven services.

. . .

Sixty years ago, writing about marketing myopia, Theodore Levitt encouraged firms to routinely ask themselves: What business are we *really* in? He asked that question in an era of products, value chains, and industries, but it remains worth asking today. To avoid digital myopia and the five traps discussed in this article, firms must answer that question in ways that are pertinent to a new era of data and digital ecosystems.

Adapted from hbr.org, August 8, 2022. Reprint H0726J

2

Democratizing Transformation

by Marco Iansiti and Satya Nadella

Over the past decade, Novartis has invested heavily in digital transformation. As the Swiss pharmaceutical giant moved its technology infrastructure to the cloud and invested in data platforms and data integration, it recruited AI specialists and data scientists to build machine-learning models and deploy them throughout the firm. But even as the technical teams grew, managers from across the business—sales, supply chain, HR, finance, and marketing—weren't embracing the newly available information, nor were they thinking much about how data could enhance their teams' work. At the same time, the data scientists had little visibility into the business units and could not easily integrate data into day-to-day operations. As a result, the investments resulted in only occasional successes (in some aspects of the R&D process, for example) while many pilots and projects sputtered.

More recently, however, pilots targeting both R&D and marketing personalization started showing business value and captured the attention and imagination of some of Novartis's more creative business executives. They became increasingly excited about opportunities to deploy AI in various parts of the company and began to earnestly champion the efforts. (Disclosure: We have both worked with Novartis and other companies mentioned in this article in a variety of ways, including board membership, research, and consulting.) They realized that technologists and data scientists alone couldn't bring about the kind of wholesale innovation the business needed, so they began pairing data scientists with business employees who had insight into where improvements in efficiency and performance were needed.

Novartis also invested in training frontline business employees to use data themselves to drive innovation. A growing number of teams adopted agile methods to address all kinds of opportunities. The intensity and impact of transformation thus accelerated rapidly, driving a range of innovation initiatives, including digitally enabling sales and sales forecasting, reconceiving the order and replenishment system for health-care-services customers, and revamping prescription-fulfillment systems and processes.

The progress in digital transformation became invaluable as the company dealt with the initial chaos of the pandemic. Novartis business teams partnered with data scientists to devise models to manage supply-chain disruptions, predict shortages of critical supplies, and enable quick changes to product mix and pricing policies. They also developed analytics to identify patients who were at risk because they were putting off doctor visits. As the Covid crisis wore on, the value of AI became obvious to managers companywide.

Idea in Brief

The Problem

Many companies struggle to reap the benefits of investments in digital transformation, while others see enormous gains. What do successful companies do differently?

The Journey

This article describes the five stages of digital transformation, from the traditional stage, where digital and technology are the province of the IT department, through to the platform stage, where a comprehensive software foundation enables the rapid deployment of AI-based applications.

The Ideal

The ideal is the native stage, whose hallmarks are an operating architecture designed to deploy AI at scale across a huge, distributed spectrum of applications; a core of experts; broadly accessible, easy-to-use tools; and investment in training and capability-building among large groups of businesspeople.

Before this wave of AI adoption, Novartis's investments in technology consisted almost entirely of packaged enterprise applications, usually implemented by the IT department with the guidance of external consultants, vendors, or systems integrators. But to build companywide digital capability, under the leadership of then chief digital officer Bertrand Bodson, Novartis not only developed new capabilities in data science but also started to democratize access to data and technology well outside traditional tech silos. The company is now training employees at all levels and in all functions to identify and capitalize on opportunities for incorporating data and technology to improve their work. In 2021, the Novartis yearly AI summit was attended by thousands of employees.

The potential for employee-driven digital innovation is impossible to calculate, but according to the market research firm IDC's Worldwide IT Industry 2020 Predictions report, enterprises across the global economy will need to create some 500 million new digital solutions by 2023—more than the total number created over the past 40 years. This cannot be accomplished by small groups of technologists and data scientists walled off in organizational silos. It will require much larger and more-diverse groups of employees—executives, managers, and frontline workers—coming together to rethink how every aspect of the business should operate. Our research sheds light on how to do that.

The Success Drivers

When we started our research, we wanted to understand why many companies struggle to reap the benefits of investments in digital transformation while others see enormous gains. What do successful companies do differently?

We looked at 150 companies in manufacturing, health care, consumer products, financial services, aerospace, and pharma/biotech, including a representative sample of the largest firms in each sector. Some were failing to move the needle, but many had made dramatic progress. Perhaps surprisingly, we found that outcomes did not depend on the relative size of IT budgets. Nor were the success stories confined to "born digital" organizations. Legacy giants such as Unilever, Fidelity, and Starbucks (where one of us, Satya, is on the board)—not to mention Novartis—had managed to create a digital innovation mindset and culture.

Our research shows that to enable transformation at scale, companies must create synergy in three areas:

Capabilities

Successful transformation efforts require that companies develop digital and data skills in employees outside traditional technology functions. These capabilities alone, however, are not sufficient to deliver the full benefits of transformation; organizations must also invest in developing process agility and, more broadly, a culture that encourages widespread, frequent experimentation.

Technology

Of course, investment in the right technologies is important, especially in the elements of an AI stack: data platform technology, data engineering, machine-learning algorithms, and algorithm-deployment technology. Companies must ensure that the technology deployed is easy to use and accessible to the many nontechnical employees participating in innovation efforts.

Architecture

Investment in organizational and technical architecture is necessary to ensure that human capabilities and technology can work in synergy to drive innovation. That requires an architecture—for both technology and the organization—that supports the sharing, integration, and normalization of data (for example, making data definitions and characteristics consistent) across traditionally isolated silos. This is the only real,

Digital transformation pays off

We studied 150 companies in a range of industries and found that revenue growth and compound annual growth rate among the leaders (the top quartile) in tech intensity were more than double those of the laggards (the bottom quartile).

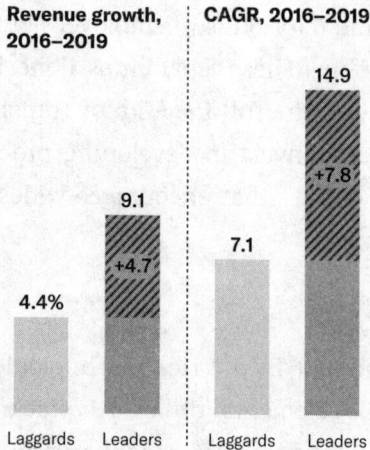

Revenue growth, 2016–2019		CAGR, 2016–2019	
Laggards	Leaders	Laggards	Leaders
4.4%	9.1 (+4.7)	7.1	14.9 (+7.8)

Source: Keystone

scalable way to assemble the necessary technological and data assets so that they are available to a distributed workforce.

Many large companies are making headway in each of these areas. But even leading companies tend to underestimate the importance of getting employees to pull transformation into their functions and their work rather than having central technology groups and consultants push the changes out to the business. As Eric von Hippel of MIT has advocated for many years, frontline users, who are closest to the use cases and best positioned to develop solutions that fit their needs, must take a central role, joining agile teams that dynamically coalesce and dissolve on the basis of business needs.

Building Tech Intensity

Our research unpacks how capabilities, technology, and architecture work together to build what we call *tech intensity*. Derived from the economics concept of intensive margin—how much a resource is utilized or applied—tech intensity refers to the extent to which employees put technology to use to drive digital innovation and achieve business outcomes. Our research found that companies that made good investments in technology and made tools accessible to a broad community of data- and tech-skilled employees achieved higher tech intensity—and superior performance. Companies that failed to develop tech- and data-related capabilities in their employees and offered only limited access to technology were left behind.

We ranked the tech intensity of the 150 firms in our study and found that the top quartile of the sample grew their revenues more than twice as fast as the bottom quartile. (See the exhibit "Digital transformation pays off.") We also found that technology, capability, and architecture indices correlated with other measures of performance, from productivity and profits to growth in enterprise value. Using an econometric technique known as *instrumental variables*, we also found evidence that the relationship between tech intensity and performance was causal: That is, greater intensity (especially investments in technical and organizational architecture) powered higher revenue growth.

Staging the Transformation

Our analysis confirms that just spending money on technology does not result in more growth or better performance; in fact, in some cases it can actually damage the business if it accentuates

The Elements of Tech Intensity

To enable transformation, companies must create synergy in three key areas:

Capabilities

- Organizational culture
- Training and development
- Low-code/no-code tools
- Agile teams
- Organizational architecture
- Citizen developers
- Product management

Technology

- Machine learning
- Deep learning
- DevOps pipelines
- Data encryption
- Real-time analytics

Architecture

- Data platform
- Horizontal integration and normalization
- Data documentation
- API strategy
- Experimentation and risk
- Data governance

divisions and inconsistencies across groups. Instead, it is the architectural, managerial, and organizational approaches to transformation that best explain the substantial and enduring differences among firms. We found that companies typically progress through five stages on their transformation journey. (See the exhibit "The stages of digital maturity.")

Traditional model

Not surprisingly, many companies fit what we consider to be the traditional model of digital innovation, whereby digital and technology investments are the province of the IT department (or other technical specialist groups) and impact is scattered across groups, mostly in inconsistent ways. IT works with business units to fund projects and manage implementation—say, for the deployment of an enterprise application or a data platform technology. The projects and their implementations are customized to the specific requirements of the individual silos, business units, or functions. The result is that over time, the technology and data infrastructure reflect the quirks of individual groups, without any consistency and connectivity. This sort of disjointed approach makes it virtually impossible to share, scale, or distribute innovation efforts across the organization.

Many businesses in the traditional model still spend a great deal of money on information technology. Consider a financial services firm we studied, whose tech and analytics budget is among the top in its industry, in both absolute and relative terms. The company has spent heavily on state-of-the-art data-platform technology and hired thousands of IT specialists and data scientists, who sit isolated in a separate IT group, while few (if any) employees on the business side are involved in the organization's digital innovation efforts. The company thus lacks the

The stages of digital maturity

Digital maturity is made up of these characteristics of organizational structure, process, tech architecture, and tech deployment. How does your company stack up?

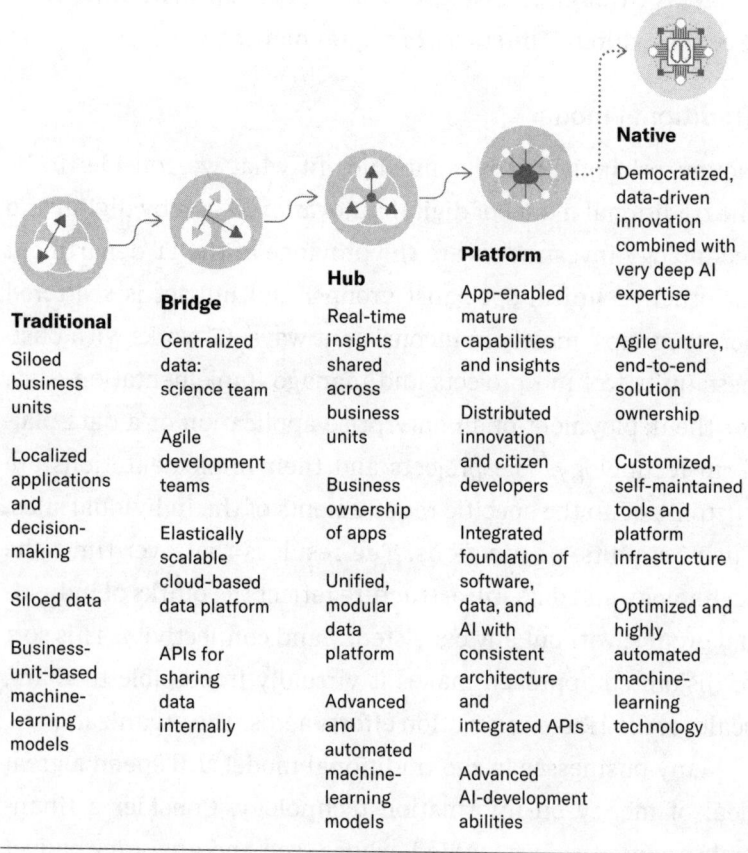

Traditional	Bridge	Hub	Platform	Native
Siloed business units	Centralized data: science team	Real-time insights shared across business units	App-enabled mature capabilities and insights	Democratized, data-driven innovation combined with very deep AI expertise
Localized applications and decision-making	Agile development teams	Business ownership of apps	Distributed innovation and citizen developers	Agile culture, end-to-end solution ownership
Siloed data	Elastically scalable cloud-based data platform	Unified, modular data platform	Integrated foundation of software, data, and AI with consistent architecture and integrated APIs	Customized, self-maintained tools and platform infrastructure
Business-unit-based machine-learning models	APIs for sharing data internally	Advanced and automated machine-learning models	Advanced AI-development abilities	Optimized and highly automated machine-learning technology

architecture and capabilities required to foster any intensity in tech adoption. Not surprisingly, the firm's IT and data sciences efforts have stalled, and business impact has been minimal.

A telltale sign that a company is in the traditional stage is that perceptions of impact among technology and business employees are dramatically different. The former perceive impact to be high (as measured by the effort they put into their work), while the latter measure it as much lower (according to how their everyday activities have benefited).

Bridge model

To break free of the traditional constraints of silos—organizational and infrastructural—companies typically start by launching pilots that bridge previously separate groups and developing shareable data and technology assets to enable new innovations. They might first focus on specific functional opportunities such as optimizing advertising, manufacturing, or supply-chain capabilities. These companies are piloting not only technology but also a fundamentally different model of innovation in which executives, managers, and frontline workers from the business side work in collaboration with IT and data scientists. Victor Bulto, Novartis's head of U.S. pharmaceuticals, was instrumental in launching early pilots (focusing, for example, on identifying at-risk patients) and served as a champion for many initiatives as the organization moved through the bridge stage. Lori Beer, JPMorgan Chase's global CIO, likes to talk about the demonstrated impact of piloting AI to simplify expense reporting and approval—a process-improvement pilot that won over many employees.

Hubs

As more and more pilots demonstrate the success of the new approach, organizations form data and capability hubs and gradually develop the capacity to link and engage additional functions and business units in pursuit of opportunities for transformation. As they progress down this path, leaders begin to realize that the bottleneck in innovation has shifted from

Digital maturity by industry

We looked at 150 companies in a range of industries and plotted the average levels of technology capability and technology architecture for each industry. Companies in consumer packaged goods, for example, tended to be at the early stage of the transformation journey; aerospace and health-care firms were much more advanced.

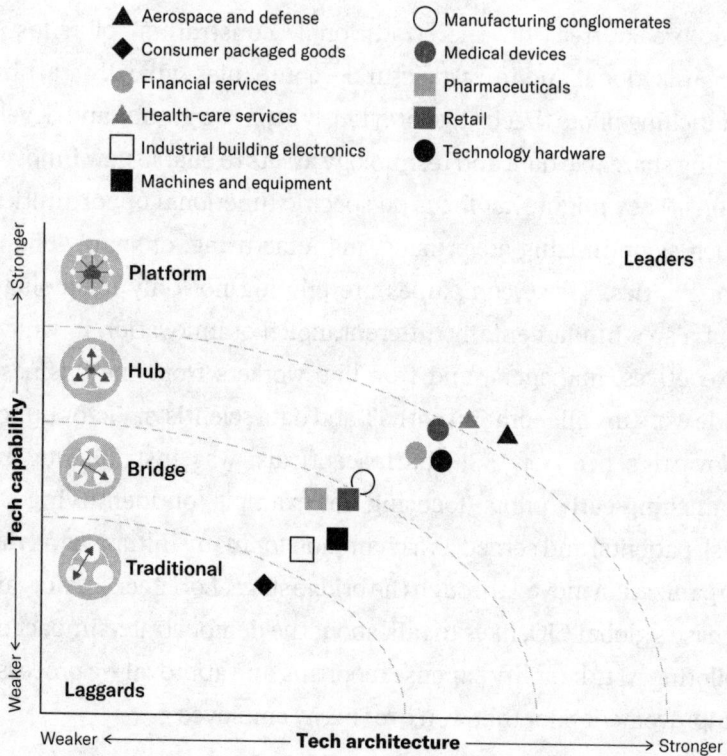

▲ Aerospace and defense
◆ Consumer packaged goods
● Financial services
▲ Health-care services
☐ Industrial building electronics
■ Machines and equipment

○ Manufacturing conglomerates
◑ Medical devices
■ Pharmaceuticals
■ Retail
● Technology hardware

Source: Keystone

investments in technology to investments in the workforce. The limiting factor at this stage is the number of business employees with the capability—the know-how and the access—to drive digital innovation. Companies thus need to invest in coaching and training a much larger community of employees.

Fidelity strives to develop what it calls *digital athletes*. It began to build hubs by creating centralized data assets (a companywide data lake, for example); now it is scaling up training for thousands of business employees, giving them the capacity to deploy digitally enabled solutions across the entire business. Digitally savvy investment specialists and tax experts, for example, are working closely with data scientists and technologists to create innovative solutions with a special focus on personalization and tailored customer impact. They've also created an app aimed at onboarding and engaging younger investors and another app for delivering AI-powered recommendations to Fidelity financial advisers, to name just a few examples.

Starbucks, too, is focused not only on technology and architecture but also on developing broad-based, agile innovation skills in its employees to power its hubs. CEO Kevin Johnson explains, "We've gone from large teams working in silos to smaller, cross-functional teams [everywhere], and from evaluating every idea as pass-fail to rapid iteration." Starbucks is now a digital innovation powerhouse, with sophisticated customer apps enabling remote ordering, loyalty programs, and payment systems along with internal systems enabling AI-based labor allocation and inventory management.

Platform model

As companies enter the platform stage, data hubs merge into a comprehensive software foundation that enables the rapid deployment of AI-based applications. Firms focus on building sophisticated data-engineering capabilities and encouraging the reuse and integration of machine-learning models. Analytics-based prediction models are applied across the business, with an increasing focus on the automation of basic operational tasks. Organizations begin to function a bit more like software companies, developing comprehensive capabilities that enable product and program management and rapid experimentation.

Over the past five years, Microsoft has gone through almost every stage of this journey. Years ago, we were just as siloed as most companies, with each product-based organization segregating its own data, software, and capabilities. As we connected and normalized data from different functions and product groups, we were able to deploy integrated solutions in areas ranging from customer service to supply-chain management.

We integrated all our data in a companywide data lake, and we built what we call a *business process platform*, which provides software and analytics components that teams use to enable innovation in areas ranging from Xbox manufacturing to managing advertising spend. We also invested in training programs for nontechnical employees, cultivating data-centric and machine-learning capabilities throughout the organization.

Native model

The most successful companies among the 150 in our study have deployed an entirely different type of operating architecture, centered on integrated data assets and software libraries and designed to deploy AI at scale across a huge, distributed spectrum of applications. Its hallmarks are a core of experts; broadly accessible, easy-to-use tools; and investment in training and capability-building among large groups of businesspeople. These companies are approaching the capacity of digital natives such as Airbnb and Uber, which were purpose-built to scale companywide analytics and software-based innovation. Airbnb and Uber are certainly not perfect, but they come close to the native ideal.

At Microsoft, we still have a lot to learn, but in some parts of the organization we are starting to approach the native model. As is common in any enterprise, the progress has not been uniform. Different groups have achieved different levels of capability, but the results overall are encouraging, as we see increasingly innovative solutions to internal and customer-facing problems. Most critically, our companywide approach to understanding, protecting, and working with data has progressed by light years.

The Imperative for Leaders

The mandate for digital transformation creates a leadership imperative: Embrace transformation, and work to sustain it. Articulate a clear strategy and communicate it relentlessly. Establish an organizational architecture to evolve into as you make the myriad daily decisions that define your technology strategy. Deploy a real governance process to track the many technology projects underway, and coordinate and integrate them whenever possible. Champion agility in all business initiatives you touch and influence. And finally, break free of tradition. Train and coach your employees to understand the potential of technology and data, and release the innovators within your workforce.

This mandate extends to technology providers. Despite much investment, technologies are still too complex and are often too hard to use and deploy. We need tools and technology that make driving transformation intuitive for frontline workers while keeping data secure. Let's not forget that until recently many of us were relying on specialists in Fortran and Cobol to model business problems and even to perform basic mathematical operations. Spreadsheets brought about a revolution in mathematical modeling; we need technology providers to bring the same revolution to AI and make using a machine-learning application as easy as creating a pivot table.

Momentum is growing. But we must sustain the efforts to ensure that companies of all stripes make it across the digital divide.

Originally published in May–June 2022. Reprint S22031

3

Strategy in an Era of Abundant Expertise

by Bobby Yerramilli-Rao, John Corwin, Yang Li, and Karim R. Lakhani

AI is changing the cost and availability of expertise, and that will fundamentally alter how businesses organize and compete. At its most basic level a business can be considered a differentiated bundle of expertise organized to accomplish specific tasks. Expertise—which we define as a combination of deep theoretical knowledge and practical know-how in a specific domain—can take many forms within a business. A doctor's office requires not only a practitioner's medical knowledge to make fast and accurate patient diagnoses but also the managerial capabilities to run a practice. A software company requires expertise not only in software engineering but also in marketing, sales, operations, and finance to bring its products to market. Companies create value by applying their expertise efficiently at scale to solve problems for their customers. Typically they possess it in a variety of areas, but most differentiate themselves through their unique proficiency in just a handful of activities that are fundamental to how they create competitive advantage. Toyota's superior expertise in lean manufacturing has helped it become one of the world's leading automakers. Walmart has built superior expertise in distribution, Procter & Gamble in consumer marketing, and Nvidia in graphics processing unit (GPU) design.

The evolution of expertise defines the evolution of business. Given the unrelenting nature of competition, companies must continually improve how they deploy expertise to remain relevant. We've seen the competitive advantage of many incumbents erode when new expertise becomes critical for success in a market. In 2007 Nokia was the global leader in mobile phones, commanding 40% of the market. Its competitive advantage stemmed from expertise in hardware and a highly tuned manufacturing process, which allowed the company to achieve huge economies of scale and scope. The smartphone era that followed required other types of expertise, especially in software. Unable to build the expertise to design and foster a cohesive software ecosystem, Nokia (along with the established phone manufacturers Motorola, Sony Ericsson, and BlackBerry) quickly lost virtually all its share to device manufacturers such as Apple and Samsung, which used iOS and Android to run their phones.

Two Fundamental Forces

Remaining on the frontier of expertise in important areas is critical to any company's success. Technological progress creates two fundamental forces that complicate that challenge.

First, the overall body of expertise in the world is constantly expanding, making it harder to stay at the leading edge in every

Idea in Brief

The Problem

As expertise becomes more widely available through AI and other digital tools, companies risk losing their strategic edge. When everyone has access to the same knowledge, differentiation becomes harder.

The Solution

To stand out, firms must go beyond simply acquiring expertise—they must integrate it in ways that align with their unique context, values, and long-term vision. This means cultivating internal judgment, fostering original thinking, and using external knowledge not as a crutch, but as a catalyst for distinctive strategy.

The Payoff

By creating a strategy that's not only informed by the best available knowledge but also shaped by the company's unique identity, leaders can make their companies more resilient, more relevant, and more difficult for others to copy.

relevant area. For example, biotech companies are increasingly leveraging AI for drug discovery—using it to analyze potential biological targets for new drugs, to design new molecules, and to predict new drug-target interactions. The field is advancing rapidly, and the number of academic papers referencing AI's role and its application in drug discovery and other areas of pharmaceutical research is growing exponentially. In 2001 fewer than 200 such papers were published. Twenty years later more than 45,000 academic biology papers referenced AI. Staying abreast of advances and insights is no longer possible for any individual scientist or biotech company looking to use AI to accelerate discovery.

Second, the cost of accessing expertise is constantly falling. Although that can benefit existing companies, it can also lower the barriers for new entrants. Think about how creator tools have transformed the media landscape by reducing the cost of accessing expertise in making and sharing high-quality content. Instagram and TikTok, for example, provide video- and photo-editing tools, audio and music integration, and analytics tools that enable amateurs to produce professional content quickly and cheaply. Large brands have embraced those platforms to reach their audiences, and individual artists and influencers have adopted those tools to launch new businesses.

We believe that the interplay between these two factors—the increasing amount of expertise required to create value and the decreasing cost of accessing that expertise—shapes companies and affects the scope of their operations. The economist Ronald Coase, in his 1937 paper "The Nature of the Firm," argued that a company's size and scope are determined by the relationship between internal and external costs. If internal costs fall, companies can expand their internal operations. If external costs fall, they will find it more efficient to source services from providers.

For most of industrial history Coase's theory predicted the evolution of businesses as the cost of access to expertise fell. During the Industrial Revolution mechanization led to process standardization and labor specialization, dramatically reducing production costs. Specialized expertise in areas such as machinery operations and maintenance became more abundant and accessible, allowing enterprises to expand. To keep pace with competition, they invested heavily in growing their internal expertise across product manufacturing, finance, sales, and other functions, developing complex structures to manage sprawling operations. In recent years, however, the trend toward ever-expanding operational scope has reversed as the breadth and level of expertise needed to remain competitive has continued to grow.

Friedrich Hayek, a contemporary of Coase's, believed that markets and price systems would be more effective than companies at accessing and managing dispersed knowledge in society. Since the 1980s several technological innovations have led companies to rely more and more on markets to access expertise far broader and deeper than what could practically exist within a single entity. Those that use third-party business and technology platform services have been able to narrow the scope of their in-house expertise, allowing internal resources to focus on the areas that drive their competitive differentiation.

Developments in communications technology have played an important role in this transition. As the cost of long-distance interactions fell, companies found it increasingly viable to outsource customer service and other processes to specialists in low-cost regions. During the internet revolution technology platform companies emerged, offering businesses access to the expertise of expansive ecosystems of partners. Today cloud platform companies such as Microsoft, Google, Amazon, and Alibaba provide cost-effective and scalable infrastructure and rich software solutions to their customers, who no longer have to build custom applications and maintain large workforces to run them.

A large direct-to-consumer retailer, for example, can now rely on Shopify to build its e-commerce website, Google to advertise and connect with consumers, Stripe to process payments, Amazon to manage logistics and fulfillment, Salesforce and Workday to manage back-office applications, and Microsoft to provide secure cloud computing and AI platforms. This modern commerce technology stack is essentially an entire business platform. It allows retailers to focus their teams, management attention, and capital on the expertise that truly matters for their brands—understanding their customers and developing innovative products to suit their needs.

The same technology stack from the same providers also enables small and medium-size businesses to compete with bigger players.

What AI Will Mean for Companies

We are at an early stage in the AI era, and the technology is evolving extremely quickly. Providers are rapidly introducing AI "copilots," "bots," and "assistants" into applications to augment employees' workflows. Examples include GitHub Copilot for coding, ServiceNow Now Assist to improve productivity and efficiency, and Salesforce's Agentforce for everyday business tasks. These tools have been trained on a wide range of data sources and possess expertise in many domains.

Although the quality of expertise embedded in these tools is already relatively high, the amount of it continues to grow swiftly while the cost of accessing it decreases. (For example, the price for developers to access OpenAI's GPT-4 model from within their own applications has fallen by more than 99% over the past 18 months.) In the relatively near future more-advanced AI agents, equipped with greater capability and broader expertise, will be able to act on behalf of users with their permission.

Companies that take advantage of AI will benefit from what we call the *triple product*: more-efficient operations, more-productive workforces, and growth with a sharper vision and focus.

Cost and time savings. Companies can transform many of their business processes and achieve new levels of efficiency by empowering employees to leverage AI for discrete tasks.

Historically companies have looked to offshoring and outsourcing to reduce costs. However, they found it cost-effective only if they outsourced an entire process. Now, with AI assistants, people can access expertise for individual tasks or steps *within* it, which allows them to make improvements without having to move the entire process. The ease and low cost of handoffs to AI mean that many processes can now be run far more efficiently. In the future workers at all levels in a company may take on more-supervisory roles, approving actions and managing exceptions as AI agents increasingly handle end-to-end execution.

Coding is an early example of a process undergoing transformation. AI tools such as GitHub Copilot, Amazon CodeWhisperer, Replit, and Cursor enable developers to outsource manual, lower-value tasks, including generating basic code, populating reference articles, and suggesting unit tests to run. Developers can then focus on higher-value tasks that require judgment and creativity, such as writing nuanced code, troubleshooting, and performing security analysis, all of which lead to greater efficiencies and a better finished product. Many studies have shown that developers using various AI coding tools are able to complete tasks 20% to 55% faster. Most developers also reported that they were able to focus on more-important work and felt more fulfilled in their jobs.

AI assistants are enabling process improvements in other domains as well. A large real-world study found that generative AI assistants helped customer service reps in a call center resolve 14% more issues per hour. An experiment demonstrated that security professionals using an AI assistant completed tasks 7% more accurately and 23% faster. AI assistants are catalyzing transformation in many other processes, including company-specific ones.

Moderna exemplifies how AI assistants can transform an entire organization's operations. By integrating advanced AI tools across its business, the company has enabled its 6,000 employees to create more than 900 specialized AI assistants that perform various tasks. Those assistants are revolutionizing processes throughout the organization, from optimizing drug doses for clinical trials to drafting responses to regulatory inquiries. Tasks that once took weeks can be accomplished in minutes, and Moderna's employees can focus on higher-value activities.

Greater workforce productivity. We argue that today expertise follows a normal distribution pattern within any given population of employees: Some of them are simply more knowledgeable or skillful than others owing to experience or inherent capabilities. As companies adopt AI assistants, those assistants will effectively put at least a base amount of expertise into the hands of every employee who uses them, enabling that person to perform better. We already see a pattern in early deployments of AI assistants: They bring low performers up to levels previously considered average and boost the capabilities of high performers (albeit to a lesser extent).

A recent randomized controlled trial conducted by BCG and researchers at Harvard's Digital Data Design Institute (which Karim cofounded and chairs) provides evidence of this pattern. It found that a group of BCG consultants who used AI displayed greater productivity than a control group did. The AI-augmented consultants completed 12% more tasks, on average, and did so 25% faster. When BCG scored the quality of each consultant's output, it found that the use of AI led to improvements across the board—but especially with lower-skilled consultants, whose scores rose 43% while the scores of the highest-skilled employees rose 17%.

Using AI to augment the expertise and thus the capabilities of employees has multiple implications. It may reduce the time and cost to onboard new hires, broaden the pool of those who can perform specific processes, and provide more flexibility in how employees are deployed to achieve outcomes. It also has implications for organizational structure if median performance levels can be achieved with less oversight from a manager and if any employee can eventually command a range of AI agents to get jobs done. Some companies may adopt wider managerial spans, while others may choose to operate with more tightly scoped teams, with each team member managing a fleet of AI agents.

More investment in activities that matter. As AI agents and bots transform business processes and empower workforces, companies will be able to fundamentally rethink how they deploy their resources. Smart ones will identify the handful of processes in which they can provide world-class expertise and capabilities and reallocate resources to deepen the moats around those processes. At the same time, they will reduce employees' focus on noncore processes by leveraging AI-enabled platforms provided by third parties.

An early example of this shift can be seen at FocusFuel, a purveyor of caffeinated gummies. Founded in 2023 by a trio of entrepreneurs and marketers in partnership with a collective of gamers, athletes, and content creators, the company has utilized gen AI technologies across its entire value chain. The founders identified their core competencies as understanding the needs of their target market and developing innovative products. They then strategically deployed AI assistants to handle noncore activities such as market analysis, supplier identification, packaging design, and marketing strategy. By building on its

AI-enabled platforms, FocusFuel was able to set up its entire operation in just a few months, efficiently outsourcing manufacturing and distribution to third-party experts. That approach allowed the founders to focus their time and resources on refining their product strategy and building relationships with their customer base—areas where their unique expertise provides a sustainable competitive advantage. FocusFuel's rapid launch, lean operating model, and early growth (the company says revenues "reached seven figures" in its first eight months) suggest that enterprises can thrive by strategically leveraging AI platforms for noncore functions while concentrating their efforts on differentiated, value-creating activities.

Getting Going

Clearly the companies that are best at continually increasing their triple-product return will have the greatest chance of competitive success. But getting there is hard. It involves meeting digital transformation requirements, aligning teams around a new course of action, helping people across the company change their behavior to maximize the benefits of working with AI, and reallocating budgets. So how should companies proceed?

Let's start with the assumption that organizations should be far along on the digital transformation journey, especially with the digitization of data, the adoption of cloud computing, and the establishment of security and governance protocols. Digital transformation can be a long process; expecting to complete it before trying to realize any benefits from AI is unrealistic. It's best to begin with a small number of business processes, maybe even just one, in which AI can be easily deployed and where it has already proved valuable for other companies. Coding, customer service, marketing, and general productivity tasks are a few prime examples. Those processes usually have digitized data and software-centric workflows that make achieving AI-based improvements easier.

AI safety and governance are foundational to success as well. Given the risks of bias, misinformation, deep fakes, and cyberattacks, companies have to establish clear guidelines and principles to safeguard their AI efforts. Again, these are hard to create in the abstract and much easier when a specific business process is the focus. Trust and safety should be among the top priorities for any organization exploring artificial intelligence.

As with any transformation, persuading people and achieving organizational change can be even more challenging than technical implementation. In addition to the tried-and-true principles of change management, an effective way to drive change in this case is to cultivate a group of employees who are early adopters of AI in select processes and empower them to become AI champions within the organization. They can serve as role models and peer mentors, accelerating adoption at every level of the company.

However, the imperative that all employees learn AI cannot be overstated. Moderna's success in rapidly deploying gen AI across its enterprise was enabled by the creation of an "AI academy"—a mandatory internal course that included 20 hours of training on AI and its business impact. Research shows that AI can fail at the front lines, and employee training is an important way to avoid that outcome.

Finally, a budget should be allocated for even early activities. As companies realize a greater triple-product return, the cost of using AI will be more than offset by productivity gains. Organizations will begin to regard AI as a core element in the budgets of all divisions, be they lines of business, functions, or the corporate center.

Coursera provides a great example of a company that started with a specific process—in this case coding—and now innovates rapidly and broadly with AI across a range of activities. When ChatGPT was introduced, Coursera's CEO, Jeff Maggioncalda, quickly recognized the potential of gen AI and started bringing it into his company. He believed that Coursera's software engineers needed to better understand emerging gen AI capabilities to truly create value. His company hired a firm that specialized in gen AI coding techniques to conduct training for its software developers. It also introduced an AI assistant that helped its engineers code more efficiently. Equipped with knowledge and tools, the teams at Coursera were able within just a year to incorporate AI capabilities into several products, such as translation, personalized learning, and automated course creation. By learning and doing, Coursera's employees have positioned themselves to continue innovating and to stay ahead of the competition as new AI capabilities become available.

Implications for Strategy

If companies derive value from providing a differentiated bundle of expertise, how can they continue to be relevant when improvements in AI's core capabilities make some or all of that expertise more easily available to competitors and customers? What is the basis for value capture in an era of abundant expertise?

We believe that every company will need to reevaluate its strategy in this changing era and will have to ask itself three questions.

1. *Which aspects of the problem we now solve for customers will customers use AI to solve themselves?* Consider the work of travel agents. For many years customers have been able to find information about travel destinations and book reservations online. Now they can simply consult AI apps to create tailored travel itineraries based on their unique preferences. As AI's ability to take action improves, it will be able to make reservations as well. Travel agents will need to reinvent themselves—perhaps by organizing unique events and experiences for their clients.

2. *Which types of expertise that we currently possess will need to evolve most if we are to remain ahead of AI's capabilities?* Companies must continue to develop their unique expertise to provide value beyond what AI provides. For example, in the medical field AI can in some cases perform image-based diagnoses more accurately than physicians can. As doctors' offices enter this new age, they will need to acquire nontechnical capabilities, such as empathy, caregiving, and working collaboratively with a team of healthcare professionals to design the right course of treatment for a patient.

3. *Which assets can we build or augment to enhance our ability to stay competitive as AI advances?* As AI comes to offer a broader range of expertise, companies will need to look to other sources of durable advantage. Those that AI is currently unlikely to affect include brands, customer relationships, ownership of scarce physical assets, and network effects. For example, a consumer-product designer can now use AI to create new prototype designs to meet

certain specifications. The designs are informed by the quality and nuance of the specifications, which will have been shaped by more in-depth and insightfully designed customer research. The ability to gather that research may eventually become more differentiating than the raw ability to create designs. A deep and trusted relationship with customers could be the best way to underpin and sustain that capability.

. . .

Without question, companies will continue to use bundles of differentiated expertise and other hard-to-replicate assets to create and capture value. But the expertise and assets that proved valuable in the past will need to be reexamined as AI improves. Over time the organizations that fully exploit AI to rapidly adapt their operations and strategy are the ones that will thrive.

Originally published in March–April 2025. Reprint R2502W

How to Speed Up Your Digital Transformation

by Benjamin Mueller and Jens Lauterbach

A s digitalization proliferates across industries, many organizations are confronting the question of how to integrate fragmented and often makeshift efforts in a way that's sustainable. At the same time, for any such initiative, whether the goal is to safeguard business continuity or enable digital innovations, one of the key questions for managers is this: Are there ways to speed up digitalization and make outcomes more predictable? This is particularly pertinent for small and medium-sized organizations that need to be more targeted in their efforts and may not have the resources to engage in the "fail fast" approach often heralded by advocates of the digitalization movement.

Based on our research, we recommend three levers for accelerating digitalization projects that will help organizations of any size reap the benefits of true transformations. These levers are rooted in the idea of complexity-in-use, a concept we developed to help understand the difficulties users face when trying to cope with the impacts of new digital tools on their work. Once managers master this form of complexity, they'll be able to plan and focus their digitalization efforts and deliver more effective transformations.

Our Study

Our insights are based on a two-year research study at one of the leading banks in Europe, which replaced its core banking system. We shadowed one of the bank's business units that provides shared after-sales services connected to the bank's mortgage and loan business. In our study, we focused on the different teams across the unit's core departments, the differences in their approaches to digitalizing their work with the new system, and their success.

We conducted over 60 interviews with stakeholders at various levels of the unit and closely observed day-to-day operations, starting with employees' established work routines in using a 30-year legacy system and ending when the unit's executives felt their teams were performing well with the new system. We were particularly interested in the contrast between departments that managed to use the new system effectively and quickly and those that struggled for a prolonged period. Analyzing these struggles allowed us to identify both the underlying mechanisms that constitute complexity-in-use and the responses to it that worked.

Key Findings

Complexity-in-use explains why learning and using a digital tool is easy and straightforward for users in one context and difficult and cumbersome in another.

Idea in Brief

The Problem

Despite significant investments in digital technologies, many companies struggle to accelerate transformation. They tend to focus too narrowly on technology implementation, overlooking the broader organizational shifts required. Siloed teams, unclear priorities, and resistance to change slow progress and dilute impact.

The Solution

To move faster, companies must treat digital transformation as a strategic capability—not just a tech upgrade. This means aligning leadership around a shared vision, empowering cross-functional teams, and embedding digital goals into core business objectives. Leaders should prioritize speed over perfection, encourage experimentation, and remove barriers to execution.

The Payoff

By taking these steps, executives can lead transformations that are not just faster, but smarter, more resilient, and more deeply integrated into how the business creates and captures value.

In our study, complexity-in-use led to vastly different digitalization journeys for different departments, even though they all used the same system for their respective tasks. For example, one group of clerks used the new SAP-based loan management system to enter new contracts. For them, learning how to do their work with the new system was easy. In stark contrast, clerks who needed to make edits to loans in stock had a much harder time learning how to work with it. Clerks in the former group achieved effective use within six to eight weeks, but those in the latter group needed over six months to do their work effectively again.

We found that two dimensions explain this difference: The first, system dependency, looks at how much of a user's task is represented in the system—that is, how much of the task and the relevant environment is implemented in the system through data and algorithms. The second dimension, semantic dependency, analyzes the degree to which users need to understand how the business logic of their task is implemented in the system. Digitalized tasks (that is, tasks that are supported by a digital tool) that have a high degree of both dimensions are the most complex.

In our example, the data entry clerks' task only requires the loan contract data to be represented in the system. An understanding of the deeper logic of a loan contract is not required to enter data successfully, nor is understanding how loan contracts are represented or processed in the system. Therefore, learning the system for that specific task is relatively straightforward.

But it's a different story for the clerks editing loans. Beyond just the loan contract data, a significant number of their tasks rely on additional business concepts (for example, loan status or certain calculation rules) that are represented in the system. These clerks also need to understand what the data means and how it's being processed in order to make correct edits to the loan. In effect, learning the system is much more complex and effortful.

These examples illustrate the dimensions underpinning complexity-in-use. First, system dependency increases when more business concepts are represented in the system. Second, semantic dependency increases if a deeper understanding of these concepts and how the system processes them is required. The two dependencies complement and reinforce one another; the impact of semantic dependency will be much higher if system dependency is also high.

These dependencies confront users every time they cognitively prepare for doing a task using a system. Of course, users will learn over time once tasks become routine, but in the early stages of a digitalization project, the cognitive efforts of mapping tools and tasks to one another in order to do work effectively and efficiently are often immense.

This complexity-in-use is often overlooked in digitalization projects because those in charge think that accounting for task and system complexity independent of one another is enough. In our case, at the beginning of the transformation, tasks and processes were considered relatively stable and independent from the new system. As a result, the loan-editing clerks were unable to complete business-critical tasks for weeks, and management needed to completely reinvent its change management approach to turn the project around and overcome operational problems in the high-complexity-in-use area. It brought in more people to reduce the backlog, developed new training materials, and even changed the newly implemented system—a problem-solving technique organizations with small budgets wouldn't find easy to deploy. In the end, our study partner managed this herculean task, but it took months to get the struggling departments back on track.

Three Levers for Accelerated Digitalization

Our study provides important lessons for those seeking to push their own digitalization efforts to the next level and avoids some of the problems and expenses our study partner faced. Informed by our findings and the feedback executives provided, here are our three levers for accelerated digitalization.

First, conduct preimplementation due diligence. Develop a complexity heat map that identifies the different degrees of complexity-in-use across the organization. The table shows what you'll need to build your heat map.

Develop a complexity heat map

Follow these steps to determine tasks' relative levels of complexity-in-use across your organization and determine their placement on your heat map.

	Step	Input	Output
1	Analyze relevant processes and tasks	Process diagrams, business domain glossary, interviews, observations, etc.	Process overview, list of relevant tasks
2	Analyze features of new system	Screenshots, system documentation, user training, interview data, observations, etc.	Feature and functions catalogue
3	Map system to tasks to understand which tasks are to be digitalized	Interviews with business users or management to extract objectives for digitalization effort	Extent of system dependency for relevant tasks (determines task placement along x-axis)
4	Analyze the properties of the to-be-digitalized tasks and understand how much business logic is involved	Output of step 3	Digitalized tasks and their degree of semantic dependency (determines task placement along y-axis)
5	Draw heat map of to-be-digitalized processes and compare their complexities-in-use (low vs. high)	Output of step 4	Heat map for complexity-in-use

The first two steps reveal which tasks will depend on the new system and how the system will be used for them. This allows you to move to step three, where you'll determine where on the x-axis of your heat map individual tasks are located (see the figures). Once it's clear which tasks are system dependent, step four will reveal their degrees of system dependency (the y-axis). Once you understand where a task is located on the y-axis, you can start drawing up a heat map (step five) to illustrate the relative levels of complexity-in-use in the various tasks that are to be digitalized. Place tasks that don't use the system in the "none" box as shown in the figures.

We drew up complexity heat maps for the two areas of our example. A number of tasks that the loan-editing clerks need to do come with high complexity-in-use (top figure), an indicator that transformation efforts in this area will be higher than in the low-complexity-in-use area of data entry (bottom figure).

While drawing the maps seems like a lot of effort before the actual digitalization can start, our research shows that this approach can prevent costly mistakes early on.

Second, design a step-by-step transformation plan. This enables you to direct attention and organizational resources toward areas with relatively low complexity-in-use first. Project efforts in these quick-win areas differ considerably from high-complexity-in-use areas in terms of scope, manpower, and transformation measures. When you roll out a new system in a low-complexity-in-use area, you can set up the transformation team with lightweight project governance and just a few key people, and change management can be reduced to a minimum. Here, digitalization investments are likely to pay off quickly.

Beyond financial considerations, quick wins also have an important psychological effect. Because digitalization projects are often marathons rather than sprints—requiring gradual changes to organizational structures and culture over time—successful pilot projects in the early stages serve as guiding and motivating lighthouses, enabling a lean approach to transformation management that can be adapted and improved.

Heat map for a high-complexity area

This complexity heat map example identifies tasks in the higher-complexity-in-use area that increased transformation effort.

Example 1: Loan-editing clerks' tasks

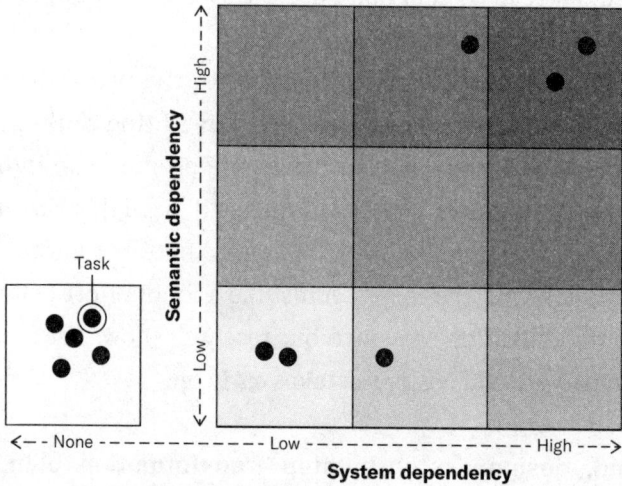

Heat map for a low-complexity area

This complexity heat map example shows tasks of the lower-complexity-in-use area that managed to transform quickly.

Example 2: Data entry clerks' tasks

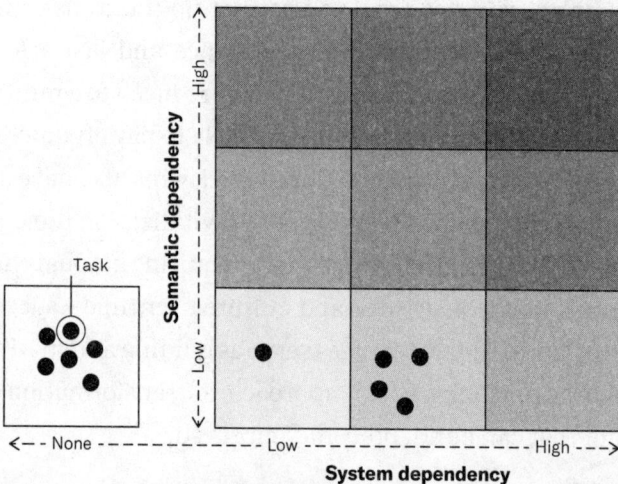

Applying the first two levers helps recoup early investments more quickly and builds momentum to carry out more complex efforts later on.

Third, develop tailor-made transformation measures. For example, low-complexity-in-use areas might only require traditional feature-based trainings to introduce employees to a new system. In contrast, other training measures are required to tackle the difficulties typical of high-complexity-in-use areas. Ongoing task-focused trainings are needed here, along with a temporary suspension of performance goals and opportunities for self- and social learning, to name just a few measures that worked in our example. Complexity heat maps help design and direct these efforts to where they're needed most because they allow executives to understand which tasks are the effort drivers in an area's digital transformation. This way, organizations can use their resources well and avoid being bogged down in turnaround mode and losing precious time.

Managerial Implications

You'll find that awareness of complexity-in-use provides valuable insights that help speed up digitalization and reveals three important implications for processes, projects, and people.

For processes, system and semantic dependencies, which are important drivers of complexity, you'll need updated ways to document and model processes. Organizations must be aware of the dependencies in an area's tasks if they want to understand where effort in transformation is created and why (our first lever).

For projects, awareness of complexity-in-use opens up new perspectives on how to phase transformation projects. This will, in turn, make transformation efforts easier to plan and execute (our second lever).

For people, our work shows that one-size-fits-all digitalization approaches don't work, and for good reason. Transformation measures need to be carefully calibrated to the complexity-in-use of different areas of the organization (our third lever). This applies to the content of the trainings (learning how to use a tool versus what the availability of a new tool means for how people do their work), the format of the trainings (lecture-style versus self-learning or social learning), and the timing of the trainings (only before going live versus throughout the weeks or months after going live, until work is done effectively again).

Taken together, being aware of complexity-in-use enables managers to apply our three levers to design transformation journeys so that their companies can reap the benefits associated with digitalization earlier.

Adapted from hbr.org, August 25, 2021. Reprint H06HZ1

4

Digital Doesn't Have to Be Disruptive

by Nathan Furr and Andrew Shipilov

Near the end of a long lunch overlooking tranquil Lake Geneva, a senior vice president at a leading global company confessed to us: "We have a dozen committees on digital transformation; we have digital transformation initiatives; we are going full steam on digital transformation . . . but no one can explain to me what it actually means."

At a very basic level, the answer is simple: The much-used term simply means adapting an organization's strategy and structure to capture opportunities enabled by digital technology. This is not a new challenge—after all, computers and software have been around for decades and have brought changes both to products and services and to how we make and deliver them. But the point the SVP was making is that it has become increasingly difficult for a company to translate that answer into an action plan. Computers today can fit in your pocket or on your wrist, and the software applications that run on them increasingly enable the automation of tasks traditionally done by humans (such as managing expenses), the virtualization of hardware, and ever more targeted product and service customization. What's more, these apps can reach people everywhere: Sensors embedded in devices and interfaces permit the real-time feed of data, allowing even more informed decision-making and machine-driven recommendations. In short, digital technology is no longer in the cordoned-off domain of IT; it is being applied to almost every part of a company's value chain. Thus it's entirely understandable that managers struggle to grasp what digital transformation actually means for them in terms of which opportunities to pursue and which initiatives to prioritize.

Faced with this reality, it's not surprising that many managers expect digital transformation to involve a radical disruption of the business, huge new investments in technology, a complete switch from physical to virtual channels, and the acquisition of tech startups. To be sure, in some cases such a paradigm shift *is* involved. But our research and work suggest that for most companies, digital transformation means something very different from outright disruption, in which the old is swept away by the new. Change is involved, and sometimes radical replacements for manufacturing processes, distribution channels, or business models are necessary; but more often than not, transformation means incremental steps to better deliver the core value proposition.

In the following pages we draw on the insights we have gathered—from interviews with more than 60 companies and from the hundreds of senior leaders with whom we have interacted while teaching—to dispel some critical myths about digital transformation and to offer executives a better understanding of how businesses need to respond to the current trends.

Idea in Brief

The Problem

Many managers believe that digital transformation involves a radical disruption of the business, new investments in technology, a complete switch from physical to virtual channels, and the acquisition of tech startups.

Why It Happens

Digital technology is being applied to almost every part of company value chains, making it difficult for managers to identify priorities.

How to Fix It

The authors dispel five critical myths about digital transformation and offer executives a better understanding of how to respond to current trends.

Myth: Digital requires radical disruption of the value proposition.

Reality: It usually means using digital tools to better serve the known customer need.

Some managers believe that to achieve a digital transformation, they must dramatically alter their company's value proposition or risk suffering a tidal wave of disruption. As a result, at the start of many digital transformations, companies aspire to be like Apple and try to find a new high-tech core product or platform that will serve brand-new customer needs. Although some might succeed, we believe that the customer needs most companies serve will look much the same as before. The challenge is to find the best way to serve those needs using digital tools. As the senior executive of Galeries Lafayette, a high-end French fashion retailer, told us, "This is another modernization.

We have been around for more than 100 years, and we have had to undergo other changes in our history, such as the arrival of hypermarkets, shopping malls, specialty chains, fast fashion, brands becoming retailers, and finally e-commerce."

The shipping container company Maersk provides a good example of what this executive meant. The costs of shipping are affected by global trade barriers and inefficiency in international supply chains. The industry also suffers from a lack of transparency. These are familiar challenges. What digital did for Maersk was provide a new way of overcoming them. The company partnered with IBM and government authorities to deploy blockchain technology for fast and secure access to end-to-end supply chain information from a single source. The technology, coupled with an ability to receive real-time sensor data, allows trustworthy cross-organization workflows, lower administrative expenses, and better risk assessments in global shipments. This shift allows Maersk to serve its core customers better. But Maersk has not been transformed into Google. It remains a company whose value proposition is providing a fast, reliable, cost-efficient shipping service—one with the potential to be more streamlined and transparent, thanks to a smart leveraging of digital technology.

Another good example is the Russian airline Aeroflot, which has transformed itself from one of the world's worst airlines into one of the best, with a Net Promoter Score that rose from 44% in 2010 to 72% in 2016 and a passenger load that grew from 64.5% in 2009 to 81.3% in 2016, according to company data. How? The airline used digital technology to significantly improve core activities: operations, reporting, passenger booking, scheduling, and customer care. Specifically, it created dashboards that provide management with an instant overview of more than 450 key performance indicators. The company also aggregates information from sensors installed on the planes, allowing visibility into aircraft performance and preventive maintenance and thereby reducing operating costs. The PR department was even able to lower its head count, because responding to journalists' inquiries about company data now requires less effort: It's all available on the dashboard. In addition, Aeroflot repurposed the digital architecture created to run the main airline to simultaneously run a low-cost carrier—something few other airlines have succeeded in doing. Once again, nothing has altered the company's raison d'être: It remains a passenger airline, selling seats on planes to many different destinations. It's just a more efficient and user-friendly one through the use of digital tools.

This is not to say that disruption doesn't occur. Make no mistake: Things are changing quickly, and companies that do nothing will be either disrupted or at a minimum outcompeted by those that transform using digital tools. But even in the classic industries where disruption strikes hardest, the story is always a little more complicated when you look below the surface. Whether you are disrupted or not always depends on the job you do for customers. If an incumbent can use digital tools to meet customers' needs better than a disruptive new entrant can, it will still prosper.

Take the taxi business. Uber's impact on taxis is one of the most frequently cited examples of digital disruption. The public remembers taxi drivers' striking around the world—notably including in Paris, our hometown—in the face of what seemed to be an existential threat to their livelihoods. But today taxi companies in Paris are thriving.

G7 is a traditional taxi company founded in 1905. It once had a reputation in Paris, as did many other taxi companies, for its drivers' rudeness. Fast-forward to the present: Like Uber, G7 has developed an app that allows customers to book a taxi. The app offers various service levels: sharing, regular cab, green (hybrid or electric), van, and VIP. You can use the app to hail a car from the curb, or you can jump into one standing at the corner, and you can pay the driver with the app using his or her four-digit code.

But G7 differs from Uber in some important ways: Its drivers are better trained, the cars are cleaner, and you can prebook a ride for exactly the time you want it, instead of in a 15-minute window. More important, although a G7 might be slightly more expensive on average than an Uber, it is vastly less expensive when you most need it: Uber imposes surge pricing, multiplying your fare twofold, threefold, or even eightfold, while G7's prices remain constant. It's clear that Uber's arrival forced traditional taxi companies to improve their service: G7 drivers now take etiquette lessons. But it's hard to argue that the advent of digital necessitated a wholesale reinvention of G7's value proposition.

Likewise, the hotel business has been among the industries most threatened by the rise of digital technologies, first from OTA (over-the-air) players like Expedia, next from platforms like Airbnb, and now from search providers like Google. When we interviewed Marriott's CEO, Arne Sorenson, about the impact of digital technologies, he didn't downplay the threat. "The digital forces are clearly very revolutionary and powerful and can be frightening at times," he said. "We are in an absolute war for who owns the customer."

Sorenson emphasized that technology would be a major factor in winning the war: "We have to make sure we are using technology to be more efficient in our operations, deliver service, and create a great loyalty digital platform, but also make sure we have a platform that is big enough and delivers value to our customers so that they book directly with us. We are not going to out-Google Google, but we want to make sure we have a community of folks who can relate to us. It must be through a digital platform. But that platform is about engaging our customers." And that is something Marriott has always done. Although it has launched platforms to compete with Airbnb and drive customers directly to its own site, it's also focusing on what it does best—delivering a great hotel and customer experience. Those who have stayed with Marriott or its sister company Starwood know they're unlikely to get the luxurious mattress and bedding these hotels are famous for at a typical Airbnb.

Understanding that digital transformation does not change the reason your business exists will help you identify the technologies you should focus on. Managers who believe that digital disruption requires wholesale reinvention of the core business end up running in a thousand directions. But if the challenge is simply to better address their customers' jobs to be done, they will most likely focus on the technologies that have the greatest effect on their customers (such as customer experience or relationship synergies) or their core capabilities (such as cost synergies). Your company, just like Maersk, Aeroflot, and G7, can probably continue to serve the same core customers even in the digital era. And the needs of those customers won't change—although digital will certainly provide a better way of catering to them.

Myth: Digital will replace physical.

Reality: It's a "both/and."

There is no doubt that digital often enables the elimination of inefficient intermediaries and costly physical infrastructure. But that doesn't mean the physical goes away entirely. In fact, as has been well documented, many retailers are finding ways to create a hybrid of physical and digital that taps into the advantages of each. And it's not just retailers—the same trend can be seen in many other consumer-facing businesses.

In retail, Galeries Lafayette provides a classic example. Despite intense competition from online stores, GL recognizes the importance of physical proximity to the customer, which only a brick-and-mortar store can offer. Both models have advantages: Physical helps build an emotional relationship with customers, while digital (especially AI) helps better understand customers' needs. Whereas in the past companies focused too much on the product and not enough on the customer, hybrid models can put the customer at the center of the business.

To ensure that it builds both an understanding of and an emotional connection with customers, the company is seamlessly blending the physical and digital worlds in its new store on the Champs-Élysées. The store will carry a curated selection of luxury items, and it will be staffed by salespeople hired for their ability to interact with visitors to the store, their expertise in fashion and style, and their facility with social media. These staffers, known as personal shoppers or personal stylists, will establish emotional relationships with their customers, making the physical store an initial customer attraction and touch point. Shoppers can then embark on digitally enabled transactions. The new technology will also help salespeople "remember" customers and their preferences and identify individualized perks that will appeal to them.

GL has already gone partway down this road at its flagship Boulevard Haussmann store, where employees are equipped with tablets. Customers come to the store having obtained—through online searches—a lot more information about some products than the salespeople have. The tablets allow employees to quickly browse the online catalogue and become equally well informed.

Shoppers value a physical store visit because they can see and feel actual products. They can reserve items online and try them out in the store without obligation. Alternatively, they can buy products online and simply pick them up in the store. In either case, salespeople must understand how to act like personal shoppers, and the product and customer data they have enable them to do so.

Many digital-first brands are converging on the same path. Bonobos, for example, which was born pure digital, now uses physical stores to let customers try on clothes. After a purchase the clothes are mailed directly from a centrally managed inventory. Warby Parker, another digital native, also now uses physical stores to create welcoming customer experiences. Like GL, these retailers are serving needs that digital meets poorly—creating emotional connections and dealing with the challenges of fitting clothing or eyewear—while using technology to leverage data and achieve cost efficiencies.

We're seeing something similar in the energy sector. Several electric utility companies in Europe have effectively combined the advantages of physical and digital in their connected home systems, which contain smart thermostats and a variety of sensors and detectors. Google and Amazon have entered the market for smart home devices, but utilities have the advantage of engineers (or selected contractors) who back the smart thermostats' value proposition—and customers trust those people to do installation, maintenance, and repair. Some of these companies enable preventive maintenance: If a sensor indicates that a heating system is about to break, the customer is alerted through the thermostat and can schedule an engineer's visit in advance. The same alert helps the engineer understand the problem before the visit and arrive with the right equipment to fix it. This seamless integration of physical and digital can significantly reduce visits and parts used while granting the customer peace of mind.

TUI UK, a travel agency, has also turned to a hybrid of physical and digital. Initially it occupied a very precarious place—its industry is broadly viewed as being disrupted. But as it embarked on a digital transformation, the company discovered that although many customers wanted to make their travel plans digitally, they also wanted to interact with people in retail locations, asking questions and becoming comfortable with complex itineraries.

Myth: Digital involves buying startups.

Reality: It involves protecting startups.

Often companies try to access new technologies or ideas by acquiring startups and then integrating them. This approach risks killing the startup's culture and chasing away the talent acquired during its creation. Smart companies prefer to build hybrid relationships with startups—strong enough to learn and find synergies but weak enough to avoid destroying the culture. So even though they may own the startups, they allow them to operate as semi-independent businesses.

Avnet, a $19 billion global technology solutions provider, is a good example. The company made two important digital acquisitions: Hackster.io, a platform that allows makers from around the world to post their ideas for new products (such as sensors to monitor city noise and pollution levels, augmented reality headsets, and baby oxygen monitors), and Dragon Innovation, a startup that helps companies bridge the gap between made-for-prototype and industrial-scale electronic products. These companies operate as semi-independent entities and interact with Avnet through Dayna Badhorn, its vice president for emerging businesses. Her role is to protect the acquired companies from the inefficiencies—such as excessive planning and slow product development cycles—of the parent organization while helping Avnet learn agility and the importance of doing quick experiments. Hackster and Dragon Innovation call her their guardian angel.

The importance of a guardian angel is underlined by Galeries Lafayette's experience with its startup accelerator, Lafayette Plug and Play, in which several big traditional retailers, including Richemont, Carrefour, Lagardère Travel, and Kiabi, are partners. Although GL executives spend a lot of time interacting with startups in the accelerator, the company struggled at first to translate such interactions into tangible projects inside GL, because no project leader was assigned to follow through. The situation has improved since GL appointed a manager to fill that role. GL does not buy startups from the accelerator (to avoid killing their innovative culture), so having someone to permanently liaise with them helps it maintain close relationships with accelerator members and implement the resulting initiatives. The other corporate members have followed suit, and their uptake of collaborations has improved as well.

In each case a guardian angel fights to take advantage of the best of both organizations, not only helping the startup hold fast to its mission (which is what motivates much of the talent to stay) but also linking it to the mission of the larger organization while protecting the startup team from all the bureaucracy and reporting that traditionally eat up company time. Meanwhile, the big company can take full advantage of the startup's ideas, processes, culture, and technology.

Myth: Digital is about technology.

Reality: It's about the customer.

Managers often think that digital transformation is primarily about technology change. Of course technology change is involved—but smart companies realize that transformation is ultimately about better serving customer needs, whether through more-effective operations, mass customization, or new offers. Because digital enables—even demands—the connection of formerly siloed activities for this purpose, the company must often reorganize both people and technology.

In practice this may mean changing structure—for example, in situations where a more agile structure is merited, creating internal squads with the capabilities and authority necessary to follow projects from beginning to end. Although a squad is a team, it differs from most big-company teams in being empowered to solve key problems quickly, as an entrepreneur would.

The credit card giant Mastercard has a systematic process for building such squads, overseen by Mastercard Labs. Employees from various functional areas can submit ideas to qualify for three stage awards: Orange Box, Red Box, and Green Box. The Orange Box gives employees a chance to explore their ideas and pitch them. Recipients of this award receive a $1,000 prepaid card and coaching to develop a presentation about solving a specific customer problem. At the Red Box stage people turn an idea into a concept: The team receives $25,000 for testing, prototype development, and research and a 90-day guide outlining the steps needed to refine the concept. The Green Box was designed to create a commercialized product from an official incubation project inside the labs. At this stage team members leave their jobs for six months to work on the project.

One major global bank, ING, teaches an important lesson about getting such squads to work in more-traditional organizational structures. It recognized that to assign the right employees to cross-company initiatives, and to keep them from staying too long on an initiative that should be cut, it needed to support these intrapreneurs in transitioning between roles. It has developed a set of internal processes called PIE: P for *protect*, meaning that employees who leave their jobs to work on a squad project can return to those jobs if the initiative fails; I for *independence*, meaning that squad members have their own resources and can make their own decisions; and E for *encouragement*, meaning that if the squad is successful, its work will be widely celebrated in the company.

Of course, it must also be OK for these squads to fail. Failures, even relatively late ones, should not jeopardize a career. As ING CEO Ralph Hamers explains, "We have to be honest about failures. We also have to be honest about all that we learned in the process and that by using a different approach, we learned these lessons in a fraction of the time it takes competitors."

There's a framing aspect as well. As the Norwegian telecom giant Telenor (for which Nathan has done consulting) makes its digital transformation, it has experimented with job definitions. Instead of designating individuals as product owners—people who oversee functions and P&L—it now calls them project *managers*, responsible for designing the customer journey. This shift encourages them to operate like mini-CEOs, externally focused on the customer problem and able to work quickly across internal boundaries to deliver a solution.

Finally, it's important to recognize that transitioning to squads can be a painful process. In a radical example of such reorganization, ING eliminated divisions and functions and instead embraced an agile organizational structure with squads tasked to deliver improved customer journeys. When it reorganized, over a weekend, all the employees were fired and had to reapply for their jobs, through the lens of the customer need they solved. With the help of these and similar initiatives, ING plans to reduce its head count in the Netherlands and Belgium by 30%–40% over a five-year period. Not all transitions will be so dramatic, but in most cases some friction is inevitable when jobs are redefined.

Myth: Digital requires overhauling legacy systems.

Reality: It's more often about incremental bridging.

Digital transformation may ultimately require radically altering back-end legacy systems, but starting with a sweeping IT overhaul comes with great risks. Smart companies find a way to quickly develop front-end applications while slowly replacing their legacy systems in a modular, agile fashion. This can be achieved by building a middleware interface to connect the front and back ends, or by allowing business units to adopt needed solutions today while IT transforms the back end in an ambidextrous manner. Over time the pieces of the legacy system can be decommissioned, but progress in meeting customer needs doesn't have to wait until then.

For example, when TUI embarked on its digital transformation, it faced a difficult challenge: Its business operations in retail, telephone, and online were geographically and operationally separate, and back-end reservations systems in the UK were 35 years old. Technology was critical for the company at the time: The rise of Expedia and other OTA channels was threatening to totally disrupt the travel agency business. In this context it was very tempting for TUI to start its digital journey with a sweeping IT overhaul. But experience suggests that attempts to replace multiple complex, mission-critical systems all at once nearly always end in disaster. Instead, in the words of Jacky Simmonds, who was part of the leadership team, "the key was to envision the ideal customer journey and then see how it could make business sense through a digital lens."

Rather than embark on a complete overhaul, TUI developed a three-year plan to replace its technology, initially working with bespoke solutions to focus on a better customer experience. The company used this time to learn from customers what they wanted in a digital world. It then connected the front-end application to the legacy back end with a middleware interface. Next it divided the back end into modular subsystems and slowly replaced them, adding front-end functionality with each step. Every time the company upgraded a component of the back end or the front end, it first tested it in one market and then iterated the prototype to improve it before working with other business units.

Although TUI decided not to roll its reservations system out more broadly, given the diversity of its markets, a coherent digital strategy allowed the markets to work together, maximizing the investment in technology. The company has enjoyed a decade of steady growth throughout its digitization of the customer journey.

The bridging role of middleware interfaces is particularly apparent in the financial services sector. In 2015 the European Parliament adopted a new Directive on Payment Services (PSD2). One of the objectives of the legislation was to enable third-party developers to build applications and services around a financial institution. If an individual is unhappy with the bank's money-transfer fees, PSD2 makes it easier for that person to use alternative services provided by a third party. Instead of waiting to change the legacy infrastructure to address the challenges of PSD2, institutions such as Deutsche Bank and the Hungary-based OTP have focused on building APIs (application programming interfaces) that allow them to connect external providers, such as TransferWise and the AI-enabled wealth adviser Wealthify, to their legacy infrastructure.

We aren't suggesting that large companies can ignore the need to update legacy systems forever. However, postponing your digital transformation until you can update them fully or all at once is dangerous. If you break the problem into modules and create a middle-layer interface, you can maintain operational stability for the core of the organization while experimenting with satisfying customer needs.

. . .

For most companies, even those truly threatened by disruption, digital transformation is not usually about a root-and-branch reimagining of the value proposition or the business model. Rather, it is about both transforming the core using digital tools *and* discovering and capturing new opportunities enabled by digital.

Each company we have described has incorporated different digital elements in its business model, and not all the changes were disruptive or intrusive. The keys to success have been a focus on customer needs, organizational flexibility, respect for incremental change, and awareness that new skills and technology must be not only acquired but also protected—something the best traditional companies have always been good at.

Originally published in July–August 2019. Reprint R1904F

5

The Secret to Successful AI-Driven Process Redesign

**by H. James Wilson and
Paul R. Daugherty**

n the late 1940s an engineer named Taiichi Ohno began
developing the Toyota Production System, basing it on the
Japanese principle of kaizen, or continuous improvement.
At Toyota it led to constant small enhancements, with key
suggestions coming from employees at all levels in manufacturing. Rather than revolutionizing its industry through bold,
innovative, and risky endeavors, Toyota chose incremental
but relentless improvement. Today it's the world's largest automaker, and the Toyota Production System continues to be a
model of how to manage processes across an enterprise. Some
notable concepts that emerged with it have enjoyed a long af-
terlife: worker empowerment, a focus on perpetual cost reduction, total quality management, just-in-time manufacturing,

root-cause analysis, data-driven processes, and automation with a human touch (*jidoka*).

As more operations become digitized, kaizen—augmented by generative AI and other advanced technologies—is once again reshaping process management. Now that features like natural-language interfaces have made gen AI accessible to nontechnical employees, it's driving both large and small process changes. With the help of AI, employees can synthesize data of all kinds, including unstructured data. They can turn once-inscrutable masses of numerical information into insight-driven workflow improvements, continuously increasing performance, reducing waste, and achieving higher levels of quality. Rather than displacing humans, as gen AI is widely presumed to do, kaizen 2.0 is moving them to the center of new machine-assisted processes and achieving a long-held aspiration of much management theory: putting business transformation in the hands of all employees.

But successfully reimagining business processes isn't as easy as asking ChatGPT to audit workflows. To get up to speed, leaders need to learn which processes are ripe for algorithm-powered redesign and understand how other companies have used gen AI to revamp them.

In this article, drawing on decades of experience providing advice to clients about technology and innovation, we'll describe how the best companies are deploying gen AI. We'll also introduce you to the future of kaizen: one in which fully autonomous agents can act independently to achieve goals, adapt strategies, analyze their environments, and complete complex tasks. However, as with all technological adoption, humans will remain the linchpin behind gen AI's success and its ability to improve business processes.

Idea in Brief

The Problem

Many organizations are eager to use AI to redesign business processes, but they often struggle to move beyond isolated pilots or incremental improvements. As a result, they fail to capture the technology's full transformative potential.

The Solution

To unlock AI's full value, companies must treat process redesign as a strategic endeavor—one that starts with empowering employees. Natural-language interfaces have made generative AI accessible to nontechnical workers, and people throughout organizations are initiating both large and small process changes. Rather than displacing workers, gen AI is putting them in the center of machine-assisted processes that are transforming creative work, scientific discovery, physical operations, and manufacturing.

The Payoff

By putting business transformation in the hands of all employees, companies can achieve step-change improvements in efficiency and accuracy, deliver better customer and employee experiences, and scale innovation.

Empowering Employees Throughout the Enterprise

Across industries from automaking to life sciences to consumer products, and across functions from R&D to manufacturing to supply chain management, gen AI is boosting employee empowerment in new ways. At Mercedes-Benz, for instance, this is happening on the shop floor, in the supply chain function, and in software design.

The company's MO360 Data Platform connects its passenger-car plants worldwide to the cloud, enhancing transparency and predictability across its production and supply chain operations and enabling the deployment of AI and analytics on a global scale. "With the MO360 Data Platform, we democratize technology and data in manufacturing," Jan Brecht, then the chief information officer of Mercedes-Benz Group, noted earlier this year. "Data is becoming everyone's business at Mercedes-Benz. Our colleagues on the shop floor have access to production and management-related real-time data. They can work with drill-down dashboards and make data-based decisions."

Using prompts in everyday language, rather than technical database queries, a production employee can ask about assembly-line bottlenecks or hard-to-notice opportunities for streamlining processes and receive data-rich insights from the AI. Such insights amplify, rather than replace, workers' ability to generate improvements based on their own experience, powers of observation, and creativity.

The platform also helps teams identify bottlenecks in the supply chain. Meanwhile, the company's software developers are using GitHub Copilot, the AI-powered assistant that turns natural-language prompts into coding suggestions. This frees them up to spend more time addressing complex process issues and integrating software development across the enterprise.

To make data use more democratic, the company is helping people across the workforce acquire new qualifications in AI. The HR department has established Turn2Learn, a program that gives frontline employees access to more than 40,000 courses on data and AI, including extensive training in skills from prompt engineering to natural-language processing. Thanks to generative AI, skilling initiatives, and digital ecosystems like the MO360 Data Platform, process change has gone from a niche technical skill to part of employees' everyday work experience at the company.

At the automaker Mahindra & Mahindra production teams can send queries to gen-AI-driven virtual assistants and receive step-by-step instructions for repairing industrial robots. That helps them quickly resolve technical issues and reduce machine downtime. Bhuwan Lodha, the head of AI at Mahindra Group, says the technology has significantly raised shop floor morale, delivering on the worker fulfillment that kaizen promises.

Redesigning Scientific Processes

In the pharmaceutical industry, gen-AI-powered synthetic data is helping workers create data-rich processes, reduce waste, speed up analysis, and strengthen quality control. Take the drug inspection process. Pharma companies rely on automated visual systems to detect product defects. Unfortunately, the systems often generate false rejects, slowing the workflow and initiating expensive do-overs. This happens because the systems need to be trained with enormous amounts of images, but for many complex defects only a limited number of images exist.

To meet the challenge, Merck uses gen AI approaches (such as generative adversarial networks and variational autoencoders) to develop synthetic defect-image data. According to Nitin Kaul, the associate director of IT architecture, the gen-AI-enhanced system has helped Merck "understand root causes of rejects, optimize processes, and reduce overall false rejects across various product lines by more than 50%."

Drug discovery is also being transformed by gen AI. Absci, a drug-development company, is now able to create and validate therapeutic *de novo* antibodies with a computer and zero-shot generative AI—in which a machine-learning model recognizes and classifies new concepts without having any labeled examples. In other words the AI designs antibodies that will bind to specific targets without using any training data on antibodies known to bind those targets. Creating them via AI instead of through trial and error could reduce the time it takes to get new biologics into the clinic from as much as six years down to 18 months, while increasing their probability of success. Waste, as kaizen has taught us, is not only a matter of materials but also a matter of time and effort.

Augmenting Creative Processes

Several leading consumer-products companies are harnessing cutting-edge AI and digital technologies to catalyze the human creativity that drives growth in the sector. At Colgate-Palmolive, employees are now using gen AI to speed up the process of devising new product formulations. Nestlé, Campbell's, and PepsiCo are reportedly using a gen AI platform that helps employees validate new product ideas and do market research. Coca-Cola is experimenting with a platform that combines the language capabilities of GPT-4 with DALL-E's ability to produce images based on text queries. The platform allows digital artists to incorporate distinctive branded elements from the company's vast archives, giving them a canvas on which they can create original artwork that will be used in billboard advertising.

Product and component design has long been a mix of art and science—combining the experience and sensibilities of a designer with the rigor of prototyping and testing. Across industries, gen AI is accelerating and transforming numerous elements of the process: creating 3D models of new ideas, suggesting modifications to designs, recommending the materials to be used, optimizing costs, rapidly creating digital prototypes, and identifying which ideas are most promising.

To what extent can such tools empower employees to improve creative processes? Consider the case of a single research engineer at NASA. Using commercially available AI software, Ryan

McClelland reinvented the design process for specialized one-off parts at NASA's Goddard Space Flight Center, in Greenbelt, Maryland. Few organizations make more one-of-a-kind components to more-exacting standards than the space agency. These components can be critical in everything from astrophysics balloon observatories to atmosphere scanners, planetary instruments, space weather monitors, space telescopes, and even the Mars Sample Return mission.

In McClelland's new process a computer-assisted-design specialist starts with a mission's requirements and draws in the surfaces where the part will connect to an instrument or a spacecraft—as well as any bolts and fittings for other hardware and electronics. The designer might also have to block out a path for a laser beam or an optical sensor. "The algorithms do need a human eye. Human intuition knows what looks right," McClelland notes. "Left to itself the algorithm can sometimes make structures too thin." Under the supervision of expert human engineers, the AI then produces complex structure designs in as little as an hour or two. In traditional mechanical design, coming up with a design and analyzing it might take a week, followed by more iterations until an expert assesses the design for manufacturability. So it can take months of work to arrive at a solution. The structures designed with AI may look a little weird, but they are two-thirds lighter and 10 times less subject to stress than components created by the traditional design process.

Animating Physical Operations

Gen AI is also transforming the ways humans interact with complex physical systems, from robots to the human body to organizations like hospitals.

Stuttgart-based Sereact, a provider of AI-based software that automates warehouse operations, has pioneered the first commercially available solution that uses the transformer technology underlying ChatGPT to enable robots to understand natural language. The robots, trained on billions of simulated images, perform "pick and pack" tasks, which typically account for 55% of warehouse costs. Called PickGPT, the technology allows human operators to simply type text commands into a chat interface; users with no technical expertise can direct and debug the system. Ralf Gulde, the CEO, calls it "the world's most accessible way of interacting with robots."

What's next? The convergence of gen AI and digital twins, already underway, provides a glimpse of a future in which continuous process improvement becomes even more democratic. Digital twins are used to model complex systems—such as jet engines, wind turbines, factories, and human hearts—and simulate their functioning with an accuracy that allows users to remotely create solutions to any problems that arise (and often before problems arise). Digital twins can be used to make production processes more efficient, improve quality, increase operational performance, and create more-robust and -resilient supply chains.

Consider how twins are used in healthcare. Some 90% of the world's top drug and healthcare laboratories already employ them in areas like preclinical drug development. Atlas Meditech has built a platform that lets surgeons practice on a virtual brain that matches the patient's brain in size, shape, and lesion position. Digital twins of hospitals are used to make day-to-day decisions about staffing, operations, and bed management. A hospital can also use its twin to stress-test the organization against future scenarios like an earthquake with mass casualties. Researchers foresee the day when digital twins will be used to deliver precision medicine, diagnose diseases, and predict health and disease outcomes.

With gen AI now poised to expand the capabilities of digital twins, including by adding natural-language interfaces, we imagine that many more healthcare workers will have the tools to adapt processes and almost instantly respond to new needs, a giant step forward for continuous improvement.

Autonomous Agents

The new AI agents take kaizen to a new level, not only offering advice but making decisions, taking action, and improving processes on their own. They range from simple chatbots to self-driving cars to robotic systems that can run complex workflows autonomously.

Consider DoNotPay, a company that aims to help consumers save money by performing a range of tasks from contesting parking tickets to canceling time-share memberships. Until recently, DoNotPay simply identified opportunities for customers to save money and encouraged them to act. But then the company integrated GPT-4 and AutoGPT into its software. The first user of these new features was DoNotPay's CEO. He gave the agent access to his financial accounts and prompted it with a concise yet complex command: Find me money. The agent discovered $81 in unnecessary subscriptions and an unusual $37 in-flight Wi-Fi fee. Then it offered to automatically cancel the subscriptions, drafted a letter to contest the Wi-Fi charges, and checked in with the CEO for review. As icing on the cake, it even drafted and sent emails that negotiated a 20% reduction in the CEO's cable and internet bill.

Traditional software is driven by precise, rule-based instructions and programmed to produce predictable outcomes. That significantly limits its ability to act autonomously. It lacks the capacity for humanlike reasoning; decisions are hard-coded and don't incorporate the nuanced judgment and flexibility characteristic of human thought. In contrast, AI agents built on top of pretrained large language models are more dynamic and adaptable because of their ability to understand language and prompts. Agents built on multimodal foundation models have even more capability because they can generalize and understand, operate across, and combine many types of information simultaneously—text, code, audio, image, and video.

Autonomous agents exhibit three other similarities to human workers in kaizen-oriented settings:

Goal-oriented behavior. People set the goals, but AI agents act independently to achieve them, adapting their strategies when necessary. To do so, an agent can work across other software platforms at a company and interact with other organizations' software and language models to execute tasks.

Logical reasoning and planning. AI agents perceive and analyze their environments. They can break complex tasks down into their component parts and use reason to figure out the best way to achieve their goals.

Long-term memory and reflection. AI agents draw on past interactions to better understand intention and context. They learn from their experience to get better at their jobs.

Across industries, many companies are now deploying AI assistants or agents with varying degrees of autonomy. Walmart uses them to help manage inventory. At Marriott International, they optimize booking processes. At Nestlé, they improve supply chain processes. ADT is building an agent that will help its millions of customers select, order, and set up their home security systems. Toyota is developing robot agents that could act as caregivers for elderly people or operate autonomously and smoothly in production processes on the factory floor. JPMorgan Chase is developing autonomous agents that will perform complex multistep tasks in the near future.

Meanwhile, technology companies, from tech giants to smaller firms and startups, are offering platforms and tools for creating autonomous agents and systems. Microsoft's AutoGen is an open-source programming framework for building such agents and coordinating them to perform tasks. It enables agents that can converse with other agents and can help create both autonomous and human-in-the-loop workflows. Meta's React is a free, open-source JavaScript library for building user interfaces, including with autonomous agents. Amazon Web Services recently introduced Amazon Q, which helps users set up semi-autonomous AI agents that perform various tasks, including writing and debugging code. Google's Vertex AI Agent Builder and Semantic Kernel open-source development kit also help you easily build AI agents and integrate the latest AI models into your code base. OpenAI's Assistants API allows users to create agents within their own applications. The San Francisco–based startup LangChain offers an open-source framework for building applications based on large language models. LlamaIndex helps users build context-augmented gen AI applications, including autonomous agents and workflows. GenWorlds provides a platform for creating environments where AI agents interact with one another to execute complex tasks. MemGPT enables agent chatbots that can learn about you and modify their own personalities over time.

Tech companies are also including autonomous agents in their product offerings. Salesforce, for instance, sells Agentforce, a completely autonomous AI agent that can understand the full context of customer messages and independently resolve a broad range of service issues without using preprogrammed scenarios the way traditional chatbots do. (Most chatbots can handle only specific queries that have been explicitly programmed into their system.)

Ecosystems of Autonomous Agents

Completing some tasks requires more than a single agent. In those cases companies may custom-build a system of agents wherein each one is expert in a specific task. Take the mortgage-underwriting process. When a human underwriter provides the instruction "Review this loan application based on our company's lending policies," one agent might extract relevant information from the application. Another agent might act as a librarian of bank policies, making them available to agents that compare the application against them. Yet another agent might generate a final report, recommending a course of action to the underwriter considering the loan. A "connector" agent might oversee and orchestrate the activity of all these agents.

Combining multiple agents and allowing them to communicate and collaborate with one another is critical to the development of AI systems that can autonomously manage end-to-end processes. Such solutions could transform entire functions like supply chain management, production, and marketing. At first glance, this might seem to be about achieving automation on a huge scale (and more fuel for sci-fi fantasists fearing world domination by AI). But the real story here is the opportunity for continuous improvement in the tradition of kaizen.

Consider an experiment conducted by researchers from Google and Stanford in 2023. They created 25 digital human avatars, each endowed with a distinctive personality and a back story, and set them loose in a simulated online world. As the avatars interacted and went about their daily "lives," they created a credible facsimile of human behavior. They made decisions based on their memory and past experiences, without real human intervention. The ability to rely on memory and to reflect on experiences and interactions like this is what enables agents to learn from one another and create ecosystems that continuously improve processes.

In another recent experiment, researchers at Stanford demonstrated that human-agent collaboration is a far more promising approach to automating complex workflows than robotic process automation (RPA), the most widely used technology. RPA is essentially a bot, hard-coded to perform a set of actions according to predefined rules. Lacking generalized reasoning and planning abilities, bots are easily tripped up by the variations and exceptions that inevitably crop up in complex processes. They're also expensive to set up and hard to maintain in the face of changing conditions.

The first workflow the researchers experimented with was revenue cycle management at a hospital. Most hospitals have departments that handle timely payment collection, patient insurance verification, prior authorization, and claims processing. A second experiment involved invoice processing at a large B2B enterprise, which was similarly complex, given the many contracts with widely varying conditions that the company dealt with. Most of the work in the two processes was still manual, despite attempts to automate it.

To overcome the limitations of RPA, the researchers employed a multimodal foundation model that learned from humans by watching video demonstrations and reading documents, greatly reducing set-up costs. The model, which encompassed a variety of agents executing different tasks, identified every step of each workflow with 93% accuracy. It leveraged its reasoning and visual-comprehension abilities to formulate plans of action, monitored itself and corrected errors, and successfully identified the completion of a workflow with 90% precision and 84% recall. Those results suggest that the model could automate entirely new categories of workflows, such as those that contain hard-to-describe steps, require complex decision-making, or are knowledge-intensive.

As it executes a workflow, the researchers' model observes the effects of its actions and can compile a database of skills that can be transferred to other workflows. Though the goal was to achieve minimal human intervention, the researchers found that human integration into processes was critical: Humans were needed to ensure alignment with overall objectives, optimize models for interactions with people, and provide training and feedback to the agents.

. . .

As the Stanford research illustrates, though AI agents act on our behalf and in concert with one another, that doesn't mean that humans are out of the loop. Success with artificial intelligence will depend as much on people as it does on technology. While employees optimize agent models for human interaction, the agents will make decisions and operate with a greater degree of autonomy. Agents will constantly improve as they gain experience, and the humans overseeing the process will continually refine their design and performance. When both employees and AI agents are empowered, continuous improvements will be generated on both sides of the human-machine equation.

Even in the face of increasing machine autonomy, processes remain human-centered. In the coming age of autonomous agents, that will be one of the keys to kaizen.

Originally published in January–February 2025. Reprint S25012

6

Developing a Digital Mindset

by Tsedal Neeley and Paul Leonardi

When Thierry Breton took over as CEO of the French IT services company Atos, in 2008, he knew that a massive and immediate digital transformation was necessary. Annual revenue had increased nearly 6% during the Great Recession, to $6.2 billion, but Atos wasn't growing as fast as its competitors were. The company suffered from siloed business and functional groups, had limited pooling of global resources, and needed more innovation throughout the company. Digital transformation was the only way forward.

But what would that look like for an IT giant? Breton began by scaling and globalizing the company, which provides online transactional services, systems integration, cybersecurity, and more. He doubled the size of the workforce to 100,000 people, hoping to fend off the competitors all around him, including digital-born startups from Silicon Valley, India, and China.

Breton also laid out a plan to integrate AI and other data-driven technology into company processes and upskill the expanding workforce.

The three-year digital-transformation plan depended on creating a culture of continuous learning and required that employees develop what we call a *digital mindset*. Breton and his team debated options for how to approach those goals. Some believed a robust training program was the only way forward; others were convinced that people learn best on the job. They eventually created the Digital Transformation Factory upskilling certification program. The initial goal was to train 35,000 technical and nontechnical employees in digital technologies and artificial intelligence.

Notably, the upskilling program was voluntary. Breton's team launched an internal marketing campaign to encourage people to learn and get certified. It also instituted a peer and management nomination system to entice employees to join the program and offered rewards for achieving benchmarks in certification. The reasoning was that if employees got certified voluntarily, they would be more likely to internalize the new digital skills and modify their work behaviors accordingly. The learning programs accommodated everyone from data scientists and highly skilled engineers to people in traditionally nontechnical functions, such as sales and marketing.

The results exceeded expectations. Within three years, more than 70,000 people completed their digital certification, in large part because employees understood that growth at the company required digital fluency. Atos was clearly on the right track. Its revenue had reached close to $13 billion by the time Breton left the company, in 2019, to become France's European Commissioner.

Idea in Brief

The Problem

Learning technological skills is essential for digital transformation, but it is not enough. Employees must be motivated to use their new skills to create new opportunities.

The Solution

They need a digital mindset: a set of attitudes and behaviors that enable them to see how data, algorithms, and AI open up new possibilities and allow them to chart a path for success in an increasingly technology-intensive world. Employees who do so are more successful in their jobs and more satisfied at work, and leaders who do so are better able to set their organizations up for success.

Maintaining Momentum

Digital transformation often encounters resistance, and missteps are inevitable. Companies do better when they focus on two areas: preparing people for a new digital organizational culture, and designing and aligning systems and processes.

What Is a Digital Mindset?

Learning new technological skills is essential for digital transformation. But it is not enough. Employees must be motivated to use their skills to create new opportunities. They need a digital mindset. Psychologists describe mindset as a way of thinking and orienting to the world that shapes how we perceive, feel, and act. A digital mindset is a set of attitudes and behaviors that enable people and organizations to see how data, algorithms, and AI open up new possibilities and to chart a path for success in a business landscape increasingly dominated by data-intensive and intelligent technologies.

Developing a digital mindset takes work, but it's worth the effort. Our experience shows that employees who do so are more successful in their jobs and have higher satisfaction at work, they are more likely to get promoted, and they develop useful skills that are portable should they decide to change jobs. Leaders who have a digital mindset are better able to set their organizations up for success and to build a resilient workforce. And companies that have one react faster to shifts in the market and are well positioned to take advantage of new business opportunities.

Like any other change initiative, digital transformation often encounters resistance, and early missteps are inevitable. In our experience, companies do best when they focus on two critical areas: (1) preparing people for a new digital organizational culture and (2) designing and aligning systems and processes. In this article, we lay out the basic principles of this enormous undertaking, drawing lessons from Philips, Moderna, and Unilever. These companies offer a road map for developing digital mindsets in existing talent pools and aligning systems and processes to capitalize on digital proficiency.

Building a Continuous-Learning Culture

The health services company Philips recently transitioned its core competency from supplying health products to providing digital solutions. To bring employees along, it needed to create a continuous-learning environment. Philips partnered with Cornerstone OnDemand, a cloud-based learning and HR software provider, to build an AI-powered infrastructure that adapts to learners' specific needs and pace. Employees can share "playlists" of tailored lessons with colleagues, just as they share playlists on music-streaming services. The platform's social media function facilitates connection between new employees and more-experienced members who can serve as mentors, fostering more-organic peer-mentor relationships than formal matching programs do.

Philips's leaders, who serve as the continuous-learning program's teachers, have emphasized the need for not only new knowledge but a cultural shift. They assume responsibility for their team members' futures, not just for managing work tasks, and they share their expertise, knowledge, and passion during training sessions. The company collects data on how employees are using the platform, measures the correlation between continuous learning and performance, and examines how various tools help employees learn, in expected or unexpected ways.

The ability to develop a digital mindset depends on the extent to which employees internalize the undertaking. Thinking about how they will interact with and use new tools and how those tools will help them attain superior performance is essential to a successful digital transformation.

Accelerating Adoption

Digital change is often radical, and it involves shifting shared values, norms, attitudes, and behaviors. That's a tall order, so it is helpful to kick things off with a bold stroke: an act that commands attention and prompts everyone in the company to understand that a new direction is required. (See "What Inexperienced Leaders Get Wrong [Hint: Management]" on hbr.org.) Examples include doing a major reorg, making an acquisition, reallocating resources, hiring a digital transformation czar who reports to the CEO, and announcing that a legacy system is being phased out.

While signaling the new order creates momentum, it isn't enough. A bold stroke must be followed by a long march, one that begins with assessing how employees feel about the plans for digital transformation. Some may be apprehensive about the unknown; others may worry about their own capacity to learn and apply the new technology and skills to their jobs. These anxieties will affect technical and nontechnical roles. Employees may also be dubious about whether the digital transformation matters—to the company and to their jobs.

When implementing radical change, managers must carefully weigh these two key dimensions: buy-in (the degree to which people believe that the change will produce benefits for them and the organization) and capacity to learn (the extent to which people are confident that they can gain sufficient literacy to pass muster). The highest levels of adoption occur when employees are motivated to develop competence because they fully buy into the transformation strategy and feel capable of helping make it a reality.

In a digital transformation, the two dimensions combine to produce the four quadrants of a matrix of responses: oppressed, frustrated, indifferent, and inspired. (See the exhibit "The adoption matrix.") In the best-case scenario, people will be in the top right quadrant, inspired by the change and believing that they have the capacity to learn digital content. Managers should assess which quadrant each of their team members falls into and then work to move individuals from one to another as needed.

Promoting buy-in

To help engage people who don't see the value in gaining digital competencies (those in the bottom quadrants), leaders must increase messaging that stresses digital transformation as a critical frontier for the company. They should launch an internal marketing campaign to help employees imagine the potential of a company powered by digital technology. Managers should encourage their team members to view themselves as important contributors to the digital organization.

Promoting confidence

After establishing buy-in, managers should focus on boosting the confidence of team members in the two left quadrants. We have found that the more experience people have with digital technologies—whether through education or employment—the more confidence they gain. Sharing stories also helps: People can build confidence vicariously, by learning about the experiences

---

The adoption matrix

Digital transformation sparks a range of responses in employees.

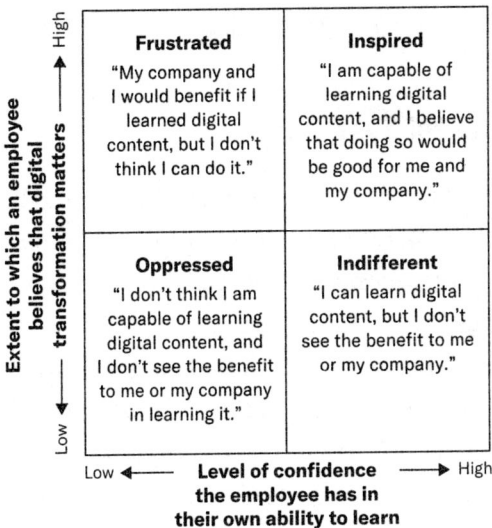

	Level of confidence → High
Frustrated	**Inspired**
"My company and I would benefit if I learned digital content, but I don't think I can do it."	"I am capable of learning digital content, and I believe that doing so would be good for me and my company."
Oppressed	**Indifferent**
"I don't think I am capable of learning digital content, and I don't see the benefit to me or my company in learning it."	"I can learn digital content, but I don't see the benefit to me or my company."

Extent to which an employee believes that digital transformation matters (Low → High)

Low ← Level of confidence the employee has in their own ability to learn → High

of peers, managers, and others. With encouragement and rein-
forcement from company leaders and direct managers, employees can begin to believe in their own capabilities. (See the sidebar
"The Elements of a Successful Employee-Training Program.")

It may seem that it would be more efficient to simply hire
people who already have the technical skills needed to bring
a workforce into the digital age. But as most companies know,
the war for talent is fierce: Hiring enough digital talent to meet
demand is nearly impossible in the current market. As a result,
recruitment must be supplemented with an expansive effort to
upskill existing talent.

Leaders should identify influencers in their ranks who have
a digital mindset and recruit them to champion the transformation and become role models for those who are reluctant. Influencers can also be very helpful in identifying areas of concern
among employees and ideas for improvements. They are likely to
understand what kind of messaging will resonate with employees. Holding training sessions about the digital transformation
and communicating new targets is also important.

Aligning Digital Systems

It is critical that organizational leaders understand how employees will deploy digital tools so that they can build technology
ecosystems and processes that foster a digital mindset and
accelerate digital transformation.

Research by Harvard Business School professors Marco Iansiti and Karim Lakhani and colleagues identifies three of the
main components of Moderna, the digital-born biotech and
pharmaceutical company. The first, foundational layer is enormous access to data, which is the source of the company's value

in developing vaccines and other therapeutics. The second is its reliance on cloud computing—a not only cheaper but faster and more agile solution than in-house servers. The third is its capacity for building AI algorithms to perform R&D processes with an accuracy and speed that is impossible to achieve manually. As Moderna's cofounder and CEO Stéphane Bancel told the scholars, Moderna is a "tech company that happens to do biology."

Historically, large pharmaceutical companies have operated in globally distributed, siloed units, but Moderna has a fully integrated structure in which data flows freely so that different teams can work together in real time. As Juan Andres, the company's chief technical operations and quality officer, has pointed out, "What's more important than having sophisticated digital tools or algorithms is integration at all levels. The way things come together is what matters about technology, not the technology itself."

In January 2020, when Moderna faced the urgent task of developing a vaccine for Covid-19, it was able to accelerate the process because integration at all levels was already in place. Bancel had hired Marcello Damiani five years earlier to oversee digital and operational excellence, and Bancel was careful not to separate the two roles. "Enabling Marcello to design the processes was key," he explains. "Digitization only makes sense once the processes are done. If you have crappy analog processes, you'll get crappy digital processes." Fully integrated systems and processes allowed Moderna employees to deploy existing digital solutions for the vaccine and build many others in-house, either designing algorithms from scratch or tweaking existing ones to perform deeper and more-specialized analyses. Only a few months after the Covid-19 outbreak, Moderna had developed some 20 algorithms for vaccine and therapeutic development and was working on many others.

Unilever, the consumer goods giant, has also adapted its sprawling global business for the digital age. For this manufacturer and retailer of household staples—more than 400 brands sold in 190 countries—success is a delicate balancing act between the specificities of local markets and the broad scale of global operations. The solution was agile teams, which could focus on customizing products to the "last mile" while simultaneously aligning their work across multiple countries using the company's digital capacities. Rahul Welde, Unilever's executive vice president for digital transformation and a 30-year veteran of the company, designed an agile-team structure that allowed members to remain globally distributed while making strategic use of data for tailored initiatives within rapidly changing local markets.

Under Welde's leadership, Unilever formed 300 10-person agile teams that were remote and global and could operate at scale. According to Welde, the strategy had three parts. The first was using enabling technology and tools, which could reduce global-local divides. With digital platforms, brands could engage directly with customers in local markets on a vastly larger scale. The second was redesigning existing processes to adapt to new technology and tools. The third was about making sure that people had access to the technology and both the skills and the motivation to use it.

Who Selects Digital Tools?

Managers and business leaders must be heavily involved in selecting and implementing digital tools. To do that, they must understand what IT departments today can and cannot

The Elements of a Successful Employee-Training Program

Continuous learning marks a new paradigm for education and career growth: The days when employees had one job and a fixed skill set for a whole career are gone. Companies that successfully upskill their workforce follow six practices.

1. Set a companywide goal for training.

2. Design learning opportunities that include all functional roles.

3. Prioritize virtual delivery, making learning scalable and accessible to everyone.

4. Motivate people to learn through campaigns, awards, and incentives.

5. Make sure managers understand the offerings so that they can effectively guide and inspire employees.

6. Encourage employees to participate in projects with digital components and hands-on learning opportunities.

do. Historically technology groups have been well equipped to handle large, enterprisewide implementations of software and to make sure that the software undergirding a company's operations is maintained and works the way it should. That remains a key function of IT for implementations of bespoke tools or ERP systems. However, most of the technologies that companies adopt to enable digital transformation are cloud-based (SaaS). Teams can simply buy licenses, download the software, and get started without ever looping in IT.

Whereas IT is accustomed to managing support applications, business leaders are best suited to the task of defining new roles and routines—and effectively reshaping organizational culture and goals. They should begin by identifying which local activities will most effectively drive larger organizational goals, as this will inform the choice of digital tools and the direction of the transformation. As technology-driven process changes lead to new roles and responsibilities, new collaborative networks will open within the organization. These networks are the real positive drivers for the organization.

The company must continually gather data to monitor the transformation effort and assess whether employee behaviors are helping or hindering what we call the *work digitization process*. Leaders should study how information flows within the organization and remove institutional obstacles that might prevent employees from adopting the new process.

Change as a Constant

According to change management theory, organizations move from a current state to a transitional state and then on to a future state. The transitional state is typically considered to be a fixed period of time in which an organization shifts from familiar structures, processes, and cultural norms to new ones. People understandably experience strong emotions during the transition, because it requires them to make sense of new perspectives and ways of behaving. During this temporary state of ambiguity, everyone's task is to negotiate between the organization's past and its future.

In a digitally driven world, however, there is no end point to the transitional phase: Digital tools change constantly and rapidly, as do the knowledge and skills necessary to use them. Organizational structures must be continually tuned to make use of new data insights, and leaders must keep working to bring employees along as the organization evolves.

Reconceiving of change as a constant process of transitioning rather than an activity that occurs between states helped Thierry Breton lead a successful digital transformation at Atos. It may be surprising that an IT company needed help with its own digital transformation, but that just underscores our point about how essential it is to cultivate a digital mindset. Just because employees have mastered one technology doesn't mean they are ready to adapt to the next one. Leaders need to view digital change as a state of constant transition that requires everyone to embrace the dynamism and uncertainty of permanent instability.

. . .

Digital technology and its impact on organizational structures, job roles, people's competencies, and customer needs is ever changing. A leader's task is not simply to adapt; it is to be adaptive. Digital transformation is not a goal that one achieves; it is the means to achieve one's unique goals. With a digital mindset, employees across the organization are equipped to seize the opportunities our dynamic world presents.

Originally published in May–June 2022. Reprint S22032

Is Your C-Suite Equipped to Lead a Digital Transformation?

by J. Yo-Jud Cheng, Cassandra Frangos, and Boris Groysberg

Historically, success rates for digital transformation efforts are dismally low. Many organizations rush to boost head count and budget, hiring teams of talented engineers, data scientists, and cybersecurity experts. But to truly succeed, transformation also needs to happen at the very top—with the individuals who set strategy and allocate resources.

Take Domino's, for example. In a mature and competitive industry, the company raised its stock price from $3 in 2008 to a high of $433 in 2020 because an integrated, digitally savvy top management team created a strategy that used data-driven experiments and decisions to redesign delivery routing systems, integrate ordering systems into a myriad of platforms (including text and smart TVs), and modernize every facet of the company.

In our experience, long-held processes and norms for selecting top executives are notoriously slow to change. Financial

Methodology

We compiled a list of all *Fortune* 1000 search assignments (across all industries) conducted at Spencer Stuart, then limited the search assignments to specific functional roles with position specifications ("specs"). We analyzed the specs for the 10 most recent assignments (spanning January 2016 through June 2020) for each role type. We searched each specs' position description, ideal experience, and key responsibilities for any mention of "technology" or "digital" in the ideal candidate's background. The use of these terms varied in specificity, ranging from broad experiences such as "experience operating in technology organizations" to more specific capabilities such as "driving digital transformations in nondigitally native organizations."

literacy is a baseline qualification for any top executive; we need to think about technological and digital literacy in the same way. These capabilities that used to be nice-to-haves are now must-haves: Companies can't afford to have an executive who might confuse discussions about the cloud with small talk about the weather.

Do today's top teams have the skills to undertake a true digital transformation? To answer this question, we conducted an analysis of more than 100 search specifications for C-suite positions in *Fortune* 1000 companies across a broad range of industries. (These listings were posted between January 2016 and June 2020. For more on the methodology, see the sidebar.)

We found that the search for tech and digital expertise has been on the rise since well before the pandemic: 59% of the searches included technological and/or digital expertise. (Tech is a broad term that typically encompasses technological

Idea in Brief

The Problem

Digital transformation is no longer optional, but many executive teams are ill-prepared to lead it. Despite bold ambitions, organizations often fall short because their C-suites lack the digital fluency and alignment needed to drive meaningful digital change.

The Solution

Companies must reconfigure their C-suites so that digital expertise is represented at the top. Senior executives must be able to foster a shared understanding of digital strategy and build a culture of collaboration and continuous learning. They should also shift from command-and-control leadership models to agile, cross-functional ones.

The Payoff

When the C-suite is digitally fluent and aligned, the organization is far more likely to execute transformation at speed and scale.

techniques, skills, systems, processes, hardware, and software. Digital can be thought of as a subset of technology, typically referring to intangibles rather than physical assets.)

Companies sought these skills across a wide variety of roles, suggesting that many had already filled key leadership roles with the right people ahead of the pandemic, but some job roles were neglected in the search for technological and digital expertise. Not surprisingly, 100% of the specs we analyzed for chief information officers, chief marketing officers, and chief technology officers sought technological and/or digital skills. But fewer than a third of the specs for chief human resources officers and chief accounting officers mentioned these skills. Falling in between—at 40% to 60%—were searches for roles such as CEO, board director, and chief financial officer. (See the table for more.)

Digital skills are not advertised equally among C-suite roles

To understand companies' talent strategies in today's increasingly digitalized world, researchers analyzed more than 100 search specifications for C-suite positions in Fortune 1000 companies posted between January 2016 and June 2020. They looked for mentions of "technology" or "digital" in the ideal candidate's background. Most companies focused their search for these skills on just a subset of job roles and were not going far enough in evolving their leadership pipelines.

	Terms mentioned in role specifications			
	Tech or digital	Tech and digital	Tech	Digital
All roles	**59%**	**28%**	**55%**	**33%**
High-focus roles				
Chief information officer	100	80	100	80
Chief marketing officer	100	70	70	100
Chief technology/digital officer	100	60	100	60
Chief customer officer	90	30	70	50
Chief sales officer/EVP/ SVP of sales	80	40	80	40
Moderate-focus roles				
Chief executive officer	60	30	60	30
President/chief operating officer	40	20	40	20
Board of directors	40	10	40	10
Chief supply chain officer	40	10	40	10
Chief legal officer/general counsel	40	10	40	10
Chief financial officer	40	0	30	10
Low-focus roles				
Chief human resources officer	30	10	30	10
Chief accounting officer	10	0	10	0

Source: Boris Groysberg et al., analysis of Spencer Stuart data

How Roles Have Evolved

Based on our study of search specs, most companies focused on just a subset of job roles in their digital transformation efforts, indicating that many had not been taking a broad enough approach in revamping their talent strategy. Successfully navigating this digital acceleration requires a shift and expansion of responsibilities across all roles throughout the organization.

Chief technology officer

In the past, CTOs were the resident experts in the opportunities and limitations that new technologies presented. Now, CTOs are being called on to lead companywide digital transformations. Not only do they need to formulate new digital strategies—identifying areas where value can be created through linkages between new technologies and the rest of the business—but equally important, they play a central role in motivating and aligning fellow employees to embrace these new initiatives and technologies. They serve as champions in building a culture that supports technology-driven strategic imperatives.

Chief marketing officer

CMOs have traditionally focused on developing and implementing their organizations' overall marketing strategy. This encompassed market research, pricing decisions, advertising strategy, and public relations. Now, CMOs must forecast future developments and adapt their strategies to contend with the ongoing evolution away from mass marketing and toward increasingly targeted and data-driven marketing. Internally, CMOs need to implement technological systems to track key performance indicators in real time, automate reports, and accelerate decision-making.

Externally, the rapid proliferation of communication platforms and online direct distribution channels makes it imperative for CMOs to ensure consistency in messaging across channels and to grapple with the constant, immediate feedback from being under the social media microscope.

Chief executive officer

The CEO has long been in charge of setting and directing company strategy, making major corporate decisions, leading the vision and culture of the organization, and serving as the public face of the company. CEOs today need to manage this hefty set of responsibilities within the context of a rapidly changing landscape. They need to develop new business targets and goals and ensure alignment between functional areas and relevant stakeholders. And when a large-scale digital transformation is needed, success will hinge on the CEO's ability to articulate the case for change, communicate a forward-looking strategy both within the firm and to external constituents, and role-model a culture that can drive the transformation.

Board of directors

Traditional responsibilities for boards include oversight of regulatory and compliance issues, monitoring financial statements, advising on company strategy, setting CEO compensation, hiring and firing the CEO, and nominating new directors. Now, the board's responsibilities are expanding to hold management accountable for a wider range of goals and targets that involve long-term innovation and investment, not just short-term financial performance. In today's world, the need for a digitally savvy board isn't just about taking advantage of future opportunities; it is also about mitigating new sources of risk. A board with directors who are technologically sophisticated will better prepare the firm for long-term success by being able to effectively advise on technological investments, select the right CEO for future needs, and manage their own learning and reskilling through new director recruitment and training sessions.

Chief financial officer

The CFO's customary responsibilities include financial forecasting, budgeting, and reporting; defining P&L investment plans; and managing shareholder activism and investor relations. Now, successful CFOs must also leverage new technologies to automate and streamline financial reporting processes and analyses. They need to maintain a progressive outlook when forecasting in an environment where revenue streams and cost structures are fundamentally changing. As the gatekeepers of much of the data needed for strategic planning and analysis, CFOs can play a central role in evaluating the costs and benefits of technological investments, and in identifying new growth areas and other opportunities for the entire firm.

Chief human resources officer

For years, standout CHROs articulated and executed a people agenda that aligned with the business strategy. Now they are also playing a key role in the future of work and future talent strategies. Although tech and digital expertise is not a primary focus in many CHRO searches, today's forward-thinking CHROs are increasingly taking HR systems online and facilitating greater employee self-service capabilities. They're implementing HR data analytics systems and finding ways to leverage data in their talent management strategy. A major challenge moving forward will be recruiting and retaining individuals with critical technological and digital skills, and managing the internal talent pipeline in light of technological advancements. At their best, CHROs serve as a partner in change initiatives and culture transformations to foster innovation and agility.

Filling the Skills Gap

This dramatic shift in job responsibilities is creating a skills gap in many companies' leadership pipelines that necessitates major changes to talent strategy. To find the right candidates, companies may need to rethink traditional promotion pathways. For example, rather than promoting the president or COO when selecting a new CEO, the search for a digitally savvy CEO might mean appointing a candidate from further down in the organizational hierarchy (a phenomenon called "CEO leapfrogging"). We've already witnessed this with Microsoft's Satya Nadella and Cisco's Chuck Robbins, who were selected, in part, for their visions of how technology will be used in the future. Nadella rose up through Microsoft's cloud and enterprise group, helping him chart Microsoft's evolution away from its traditional business products and into services. Robbins was promoted from head of sales to CEO. While sales is not typically thought of as a training ground for tech and digital expertise, this role helped him stay on top of new trends through close relationships with customers.

Job rotations can also play an important role in building a robust, future-ready leadership pipeline. Although operational roles are traditionally thought of as stepping stones to the highest corporate echelons, today's high potentials can benefit not just from rotations through jobs with P&L responsibilities, but also through tech functions. In the past, rotations into tech roles often left executives thinking, "Why am I being punished?" But IT is much more than just a support function; now, these roles can provide critical skills and experiences for leading into the future.

When the right candidates don't exist in-house, companies will need to turn to the external labor market. But they'll need to prepare to pay a premium; individuals with tech and digital skills are in demand by many companies. This trend also has implications for firms with long-standing strengths in technology, which may find themselves being raided for talent. Companies will need to take proactive steps to retain their top employees.

We see a similar skills gap emerging in boardrooms. In a global survey we conducted of board directors, more than a third of respondents indicated that they personally struggle to stay on top of risk and security issues and new technologies. Further, just 13% of boards sought technological expertise with their most recent director search. In the words of one survey respondent, this has resulted in "an overemphasis and overabundance of directors with financial and general management skills." Board turnover is typically low, and staggered terms mean that changes to board composition take a long time. Boards need to rethink ingrained norms around tenure and also spend time on training and learning opportunities for current members as they adapt to the business environment.

Preparing Yourself for Future Success

We think the next few years will see a revolution for many top management roles. The pandemic exposed the executives who were not up to the challenge of a rapid technological pivot. In some cases, it became clear that the wrong leaders were in place, and the only thing standing in the way of replacement was the lack of a suitable successor. Once boards and top teams have time to search for new candidates and more candidates are unleashed into the external market, we will likely see tremendous amounts of turnover and a very turbulent environment that will highlight the chasm between the digitally savvy haves and the traditional have-nots.

As the need for technological skills become more pressing, what can you do to stay current? We offer the following advice.

For CEOs

- You don't need to think of yourself as the expert on the leading edge. Having both the requisite leadership abilities and being at the forefront of technical proficiency is quite rare. Instead, endeavor to build the domain knowledge needed to lead organizational change, align the top team around a clear strategy, and ensure coherence across functional areas and work groups.

- Culture is an important facet of any change initiative and cannot be neglected. You play an important role in leading an innovative and creative culture that can reinforce technology-driven strategic shifts.

- Surround yourself with the right team: Identify people in the organization that have vital expertise, create a seat for them at the table, and ensure that their voices are heard. You might need to change the organizational structure to fully execute on your digital transformation.

- On the other side of the coin, you might also need to make painful decisions about switching out other roles if current executives and managers are not prepared for what's to come. One study found that 70% of digital transformation efforts do not achieve their objectives; managerial resistance is a leading reason. If leaders cannot get on board, it may be time for them to go.

For senior leaders and board members

- Always strive to stay relevant, learn, and adapt; continuously developing skills is critical to avoid falling behind.

- Consider how you can adopt a data-driven approach to *all* of the work you do. Things that have helped you succeed in the past may not be the best way any longer. Work to remain open-minded about how things are done. Resistance to new ways of working is a major barrier to companies reaching their full digital potential.

- Learn from the people who are ahead *and* behind you ("reverse mentoring"). There is much to learn from the people who are closest to new technologies and customers.

- Staying up-to-date on technological trends is critical in inspiring your team's confidence and respect for you. If you lose their trust in your capabilities, they will choose to leave or may even replace you.

For up-and-coming professionals

- Seek out development and rotation opportunities to build your existing digital and leadership capabilities, and be willing to take lateral moves to develop new ones. Even if these changes do not provide an immediate promotion, they will offer valuable experience that will round out your qualifications. (Many compelling candidates have experience in multiple disciplines, with career trajectories that are nonlinear and involve shifts and stints across different industries and functions.)

- Carefully consider the types of companies you work for. The company you are associated with will define your brand. People will make inferences about your tech savviness based on your firm. (For example, a software company looking for a new COO with software experience centered its search around executives from other software companies, frequently ruling out candidates from hardware companies even before looking at their qualifications.)

- Become an architect of your own portfolio of experiences: This could be at one company or across multiple companies. Be mindful that the career steps that mentors and other senior executives took may no longer be the way up the corporate hierarchy in today's evolving world.

The Covid-19 pandemic moved tech from the periphery into the center, and it is here to stay. Although tech and digital trends do not uniformly affect all jobs today, we believe it is only a matter of time before these skills become baseline qualifications across the board. Investing in these skills now will enhance your marketability and prepare you for the realities of our evolving business world.

Adapted from hbr.org, March 12, 2021. Reprint H06881

7

A Better Way to Put Your Data to Work

by Veeral Desai, Tim Fountaine, and Kayvaun Rowshankish

T hough every company recognizes the power of data, most struggle to unlock its full potential. The problem is that data investments must deliver near-term value and at the same time lay the groundwork for rapidly developing future uses, while data technologies evolve in unpredictable ways, new types of data emerge, and the volume of data keeps rising.

The experiences of two global companies illustrate how ineffective today's predominant data strategies are at managing those challenges. The first, a large Asia-Pacific bank, took the "big bang" approach, assuming it could accommodate the needs of every analytics development team and data end user in one fell swoop. It launched a massive program to build pipelines to extract all the data in its systems, clean it, and aggregate it in a data lake in the cloud, without taking much time up front to align its efforts with business use cases. After spending nearly three years to create a new platform, the bank found that only some users, such as those seeking raw historical data for ad hoc analysis, could easily use it. In addition, the critical architectural needs of many potential applications, such as real-time data feeds for personalized customer offerings, had been overlooked. As a result the program didn't generate much value for the firm.

The second company, a large North American bank, had individual teams tap into existing data sources and systems on their own and then piece together any additional technologies their business use cases required. The teams did create some value by solving challenges like improving customer segmentation for digital channels and enabling efficient risk reporting. But the overall result was a messy snarl of customized data pipelines that couldn't easily be repurposed. Every team had to start from scratch, which made digital transformation efforts painfully costly and slow.

So if neither a monolithic nor a grassroots data strategy works, what's the right approach?

We find that companies are most successful when they treat data like a product. When a firm develops a commercial product, it typically tries to create an offering that can address the needs of as many kinds of users as possible to maximize sales. Often that means developing a base product that can be customized for different users. Automakers do this by allowing customers to add a variety of special options—leather upholstery, tinted windows, anti-theft devices, and so on—to standard models. Likewise, digital apps often let users customize their dashboards, including personalizing the layout, color schemes, and content displayed, or offer different plans and pricing structures for different user needs.

Idea in Brief

The Problem

Although data offers enormous opportunities, most companies' strategies for realizing them are ineffective.

Why It Happens

Too often firms' data efforts fail to lay the foundations for future data uses. Individual teams create a customized data pipeline for each application that can't easily be repurposed.

The Solution

Create standard data products that can be tailored to suit the needs of various types of users and many applications. The products can be managed by dedicated teams within business units, supported by a central function that coordinates and standardizes design.

Over time companies enhance their products, adding new features (engine modifications that boost fuel economy in a car or new functionality in an app), and introduce brand-new offerings in response to user feedback, performance evaluations, and changes in the market. All the while firms seek to increase production efficiency. Wherever possible, they reuse existing processes, machinery, and components. (Automakers use a common chassis on vastly different cars, for instance, and app developers reuse blocks of code.) Treating data in much the same way helps companies balance delivering value with it today and paving the way for quickly getting more value out of it tomorrow.

In our work we've seen that companies that treat data like a product can reduce the time it takes to implement it in new use cases by as much as 90%, decrease their total ownership (technology, development, and maintenance) costs by up to 30%, and reduce their risk and data governance burden. In the pages that follow we'll describe what constitutes a data product and outline the best practices for building one.

What Is a Data Product?

A data product delivers a high-quality, ready-to-use set of data that people across an organization can easily access and apply to different business challenges. It might, for example, provide 360-degree views of customers, including all the details that a company's business units and systems collect about them: online and in-store purchasing behavior, demographic information, payment methods, their interactions with customer service, and more. Or it might provide 360-degree views of employees or a channel, like a bank's branches. Another product might enable "digital twins," using data to virtually replicate the operation of real-world assets or processes, such as critical pieces of machinery or an entire factory production line.

Because they have many applications, data products can generate impressive returns. At a large national bank, one customer data product has powered nearly 60 use cases—ranging from real-time scoring of credit risk to chatbots that answer customers' questions—across multiple channels. Those applications already provide $60 million in incremental revenue and eliminate $40 million in losses annually. And as the product is applied to new use cases, its impact will continue to grow.

Data products sit on top of existing operational data stores, such as warehouses or lakes. The teams using them don't have to waste time searching for data, processing it into the right format, and building bespoke data sets and data pipelines (which ultimately create an architectural mess and governance challenges).

Traditional data consumption versus the data product model

In the traditional approach to data solutions, use case teams identify the data they need from source systems and create datasets and feeds for only their particular solutions. That leads to a lot of replicated work and a complex data architecture that is difficult to maintain and use for new solutions.

In a data product approach, use case teams build solutions by leveraging standardized data products and wiring technologies together following consumption archetype patterns, which reduces work, simplifies the enterprise data architecture, and decreases the time it takes to realize value.

Data product approach

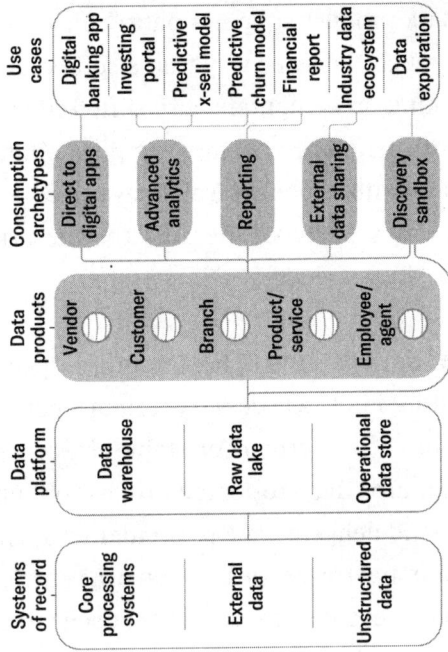

Systems of record	Data platform	Data products	Consumption archetypes	Use cases
Core processing systems	Data warehouse	Vendor	Direct to digital apps	Digital banking app
External data	Raw data lake	Customer	Advanced analytics	Investing portal
		Branch		Predictive x-sell model
Unstructured data	Operational data store	Product/service	Reporting	Predictive churn model
				Financial report
		Employee/agent	External data sharing	Industry data ecosystem
			Discovery sandbox	Data exploration

Traditional approach to using data

Flow of data ----→

Systems of record	Data platform	Use-case-specific datasets	Use-case-specific technologies	Use cases
Core processing systems	Data warehouse			Digital banking app
External data	Raw data lake			Investing portal
				Predictive x-sell model
Unstructured data	Operational data store			Predictive churn model
				Financial report
				Industry data ecosystem

Data pipelines designed for batch and real-time delivery are fragmented and duplicative.

Different technologies are employed for each use case, adding expense and complexity.

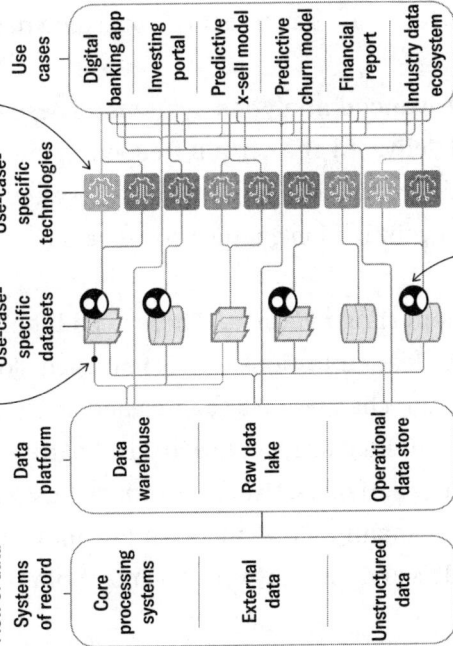

Data for each domain, such as the customer, is inefficiently reworked for every use case, and quality, definitions, and formats vary.

Each data product supports data "consumers" with varying needs, in much the same way that a software product supports users working on computers with different operating systems. These consumers are systems, not people, and our work suggests that organizations typically have five kinds. We call them "consumption archetypes" because they describe what the data is used for. They include:

1. Digital applications. These require specific data that is cleaned, stored in the necessary format—perhaps as individual messages in an event stream or a table of records in a data mart (a data storage area that is oriented to one topic, business function, or team)—and delivered at a particular frequency. For example, a digital app that tracks the location of a vehicle will need access in real time to event streams of GPS or sensor data. A marketing app designed to find trends in customer browsing behavior will need access to large volumes of web log data on demand (often referred to as "batch" data) from a data mart.

2. Advanced analytics systems. These too need data cleaned and delivered at a certain frequency, but it must be engineered to allow machine learning and AI systems, such as simulation and optimization engines, to process it.

3. Reporting systems. These need highly governed data (data with clear definitions that is managed closely for quality, security, and changes) to be aggregated at a basic level and delivered in an audited form for use in dashboards or regulatory and compliance activities. Usually, the data must be delivered in batches, but companies are increasingly moving toward self-service models and intraday updates incorporating real-time feeds.

4. Discovery sandboxes. These enable ad hoc exploratory analysis of a combination of raw and aggregated data. Data scientists and data engineers frequently use these to delve into data and uncover new potential use cases.

5. External data-sharing systems. These must adhere to stringent policies and agreements about where the data sits and how it's managed and secured. Banks use such systems to share fraud insights with one another, for example, and retailers to share data with suppliers in the hope of improving supply chains.

Each consumption archetype requires different technologies for storing, processing, and delivering data and calls for those technologies to be assembled in a specific pattern. This pattern is essentially an architectural blueprint for how the necessary technologies should fit together. For example, a pattern for a sandbox would most likely include technologies for setting up a multiuser self-service environment that can be accessed by data engineers across the company. The pattern for an advanced analytics system using real-time data feeds might include technologies for processing high volumes of unstructured data.

Like a Lego brick, a data product wired to support one or more of these consumption archetypes can be quickly snapped into any number of business applications.

Consider a mining company that created a data product providing live GPS data feeds of ore-transport-truck locations. It was designed to support all the archetypes except external data sharing for its first use case—improving ore-processing yields. The company soon discovered the product had uses far beyond that. Once it was made available more broadly in the organization, several entrepreneurial employees immediately leveraged it to eliminate bottlenecks in the mine transport system. In just three days they built a prototype of a truck-routing decision support tool that reduced queuing time and carbon emissions. If they'd had to engineer the data from scratch, it would have taken nearly three months.

As word continued to spread, employees interested in other issues that involved trucks—such as safety, maintenance, and driver scheduling—tapped into the data to find answers to thorny questions and to build revenue-generating solutions that previously would have been impossible.

Managing and Developing Data Products

Whether they're selling sedans, software, or sneakers, most companies will have internal product managers who are dedicated to researching market needs, developing road maps of product capabilities, and designing and profitably marketing the products.

Likewise, every data product should have a designated product manager who is in charge of putting together a team of experts to build, support, and improve it over time. Both the manager and the experts should be within a data utility group that sits inside a business unit. Typically, such groups include data engineers, data architects, data modelers, data platform engineers, and site reliability engineers. Embedding them within business units gives the data product teams ready access to both the business subject-matter experts and the operational, process, legal, and risk assistance they need to develop useful and compliant data products. It also connects teams directly with feedback from users, which helps them keep improving their products and identify new uses. The first release of the customer data product at the national bank, for instance, focused on customer demographic profiles and information on transactions. Subsequent releases included data on customer interactions and on prospects, attracting significantly more data users and supporting teams developing other applications. The cost savings and incremental revenue realized by the product's early uses funded the next phases, creating a sustainable business model.

A company also needs a center of excellence to support the product teams and determine standards and best practices for building data products across the organization. For example, the center should define how teams will document data provenance, audit data use, and measure data quality, and should design the consumption archetype patterns for data product teams to use. This approach can eliminate complexity and waste. In addition, the center can be a resource for specialized talent or data experts when demand for them surges within utility groups or business-use-case teams. For example, at one telecom provider we worked with, computer vision experts, who are scarce but often in demand, sit within the central hub and are deployed to business units on request.

While most companies already have some, if not all, of the talent needed to build out their utility groups and centers of excellence, many will need to deepen their bench of certain experts, particularly data engineers who can clean, transform, and aggregate data for analysis and exploration.

This was especially true for the mining company, which needed to grow its data engineering staff from three to 40 people. To fill that big gap, its leaders took a stepped approach. They hired contractors to get immediate work done and then embarked on far-reaching recruiting efforts: hosting networking events, publishing articles on LinkedIn, upgrading the skills of the software engineers already on staff, and developing internship programs with local colleges and universities. To improve retention, they created a guild for data engineers, which helped them build their skills and share best practices. The company also crafted individualized plans for data engineers that ensured those professionals had a clear growth path after joining the company.

Tracking Performance and Quality

To see whether commercial products are successes, organizations look at barometers like customer sales, retention, engagement, satisfaction, and profitability. Data products can be evaluated with commensurate metrics, such as number of active monthly users, the number of applications across the business, user satisfaction, and the return on investment for use cases.

The telecom company tracked the impact of its first data product—which provided comprehensive data on critical cellular-network equipment—in 150 use cases. They included investment decision systems, scenario-planning systems, and network optimization engines. In total they're set to produce hundreds of millions of dollars in cost savings and new revenue within three years. The company estimates that over the first 10 years the use cases will have a cumulative financial impact of $5 billion—providing a return many times over on its initial investment.

And just as manufacturers routinely use quality assurance testing or production line inspections to make certain that their products work as promised, data product managers can ensure the quality of their offerings' data. To do so they must tightly manage data definitions (outlining, say, whether customer data includes only active customers or former and prospective customers as well), availability, and access controls. They must also work closely with employees who own the data source systems or are accountable for the data's integrity. (The latter are sometimes called "data stewards.")

Quality can suffer, for instance, when the same data is captured in different ways across different systems, resulting in duplicated entries. This was a risk with the national bank's customer data product. So its product manager worked with the stewards of the company's various customer data repositories and applications to institute a unique ID for each customer. That allowed the customer data to be seamlessly integrated into any use case or with any related data product. The product manager also partnered with the center of excellence to develop the standards and policies governing customer data across the enterprise and to monitor compliance—all of which facilitated reuse of the data product while building trust among users.

Where to Start

Leaders often ask which data products and consumption archetypes will get the highest and fastest return on investment. The answer is different for every organization.

To find the right approach for their companies, executives need to assess the feasibility and potential value of use cases in each business domain (this might be a core business process, a customer or employee journey, or a function) and group them first by the data products they require and then by the consumption archetypes involved. Categorizing the use cases like this helps leaders more efficiently sequence work and get a faster return on investment. For instance, they may end up pushing some lower-value use cases ahead if they leverage the data products and consumption archetypes of higher-value use cases.

For the executives at the national bank, this approach illuminated several priorities. First they saw that a customer data product that supported their most critical fraud-management and marketing use cases could generate tremendous value. Then they identified the kinds of data the product needed to gather first. Some of those use cases called for basic customer identifiers and reference data (such as demographic or segmentation data) while others required comprehensive customer behavioral data. The bank also realized that the two consumption archetypes it should pursue first were a discovery sandbox and advanced analytics, which in combination would support most of the company's priority fraud and marketing use cases.

Data product decisions often involve trade-offs between impact, feasibility, and speed. Ideally, the initial target products and consumption archetypes will immediately apply to high-value use cases and a long pipeline of others, as the telecom provider's product for its network equipment did.

However, feasibility considerations may cause a company to adjust its approach. For example, it may make sense to build momentum first in an area of the organization that has data expertise and has gotten some traction with data products, even if that isn't where the biggest opportunity lies. We saw this happen at the mining company. It initially chose to develop two products that supported its ore-processing plant, where use cases had already been successfully proven, the managers were enthusiastic to pursue more, the team had a lot of prepared data to work with, and experts with deep expertise were available to help.

. . .

Most leaders today are making major efforts to turn data into a source of competitive advantage. But those initiatives can quickly fall flat if organizations don't ensure that the hard work they do today is reusable tomorrow. Companies that manage their data like a product will find themselves with a significant market edge in the coming years, thanks to the increases in speed and flexibility and the new opportunities that approach can unlock.

Originally published in July–August 2022. Reprint R2204G

Five Pillars for Democratizing Data

by Hippolyte Lefebvre, Christine Legner, and Elizabeth A. Teracino

B ecoming data-driven has been a North Star for most companies, and those that have not yet managed to infuse data into their organization's cultural DNA, preventing its use at scale, are now lagging behind their competitors. The disparity is growing. However, integrating and leveraging data from all corners of the organization is no easy feat; even digital natives such as Airbnb, Netflix, Uber, and the like have struggled because it isn't possible to have a data scientist at every level.

To become data-driven, companies must embrace a new management paradigm in which they empower not just data experts but everyone within their organization to work with data, irrespective of their comfort level or expertise with it. This shift can be achieved by building what we call a data democracy, the benefits of which include, but are not limited to, improved agility and expedited data-driven decision-making at all levels.

Managers often confuse the concept of democratizing data with universal access to data. It is instead about ensuring that,

Data Democratization

Data democratization is an enterprise's capability to motivate and empower a wide range of employees—not just data experts—to understand, find, access, use, and share data in a secure and compliant way.

over time, employees without "data" in their title feel comfortable enough to incorporate data into their daily activities and become "data citizens" (with rights and obligations). It entails an organizationwide cultural shift, teaching this wide range of employees with data from their own functional position or domain to contribute to business value creation and to scale data and AI for innovation.

What can senior leaders and managers learn from the digital natives and incumbents who successfully managed the transition to a data democracy? Drawing on a combination of case study research of both digital natives and digitalized incumbents, and by working directly with hundreds of *Fortune* 500 data executives and teams at various stages of building data democracies at their organizations, we present five enabling areas, or *pillars*, that we found help managers empower employees and make the transition, as well as challenges they help overcome.

1. Broaden data access by rolling out data catalogs and marketplaces

The first step is to ensure that employees have access to the data they need. While this sounds simple and even obvious, there are usually bottlenecks that need to be identified and removed first.

Idea in Brief

The Problem

Despite growing investments in data infrastructure, many organizations struggle to make data accessible and usable across teams. It remains locked in silos, underutilized by nontechnical employees, and disconnected from day-to-day decision-making.

The Solution

To truly democratize data, organizations must build around five key pillars: Broaden data access by rolling out data catalogs and marketplaces. Stimulate the generation of data-driven insights through self-service. Level up data literacy with specific curricula for personas or role families. Advance data practices by creating communities. And promote data through various corporate communication channels.

The Payoff

When these pillars are in place, companies can unlock the full value of their data—empowering employees at all levels to make smarter, faster, and more confident decisions.

For example, data is usually locked away in applications and not accessible for further use, preserving data silos.

To address these silos, employers can start by setting up a way for employees to see the data and be able to request access to it. To do this, firms typically set up dedicated platforms such as data catalogs or equivalent metadata hubs offering a library-like experience. This allows all employees to browse for data before shopping it or requesting access to it.

Digital native Airbnb successfully implemented this method. To expand access to gate-kept knowledge across fragmented data teams, one of Airbnb's first investments was to launch the data discovery and exploration tool Dataportal. The platform notably replicates the firm's data ecosystem as a graph connecting relevant data and data assets with enterprise resources, such as related teams, ongoing projects, and business outcomes. This facilitates data discovery and context retrieval by anyone in the company who may need certain access to various data. Similar data discovery initiatives have been observed at LinkedIn (Datahub), Netflix (Metacat), Uber (Databook) and Spotify (Lexikon).

Many traditional organizations have historically and in opposite fashion emphasized the compliant and secure use of data, even restricting data access entirely. Helping employees to see the data—to know it exists in the first place—with a data catalog is vital, while still allowing for governance and the compliant and secure use of data on the back end.

2. Stimulate the generation of data-driven insights through self-service

Traditionally, specialists in IT departments have generated reports and analyses, but this way of working is slow because it always incorporates middlemen. Such centralized approaches do not scale and are major barriers to data-driven innovation. A more recent solution to this challenge is self-service tools that let regular employees create their own reports or insights, helping individuals and teams form data-driven narratives from their own data and enhance decisions within or across functional areas. Some examples of such tools that have gained headway recently are Tableau, MicroStrategy, Power BI, and Alteryx.

Despite the clear advantages, these are still often only available to a very narrow audience in most firms. To move forward, companies need to offer self-service environments so that employees are able to retrieve existing analytical products (for example, a net sales dashboard) or create their own, thus stimulating the production of data-driven insights to support key business decisions across the organization. These tools additionally relieve IT and data specialists of repetitive and nonstrategic tasks.

Uber addressed this pillar by creating a self-service analytics platform that is "built for the experts, designed for the less technical people." Its goal is for data to be used to inform all decisions in the company. Employees are thus encouraged to engage in analytics projects (for example, A/B testing) in a trustful environment supported by data experts.

Airbnb took a similar approach with Knowledge Repo, which has the goal of sharing trustful and validated analytical insights across the firm. All analytical work is peer-reviewed in terms of reproducibility, quality, consumability, discoverability, and documentation before being published. Spotify's new Experimentation Platform was created so that it can adopt a truly data-informed product development process by putting together product managers, data engineers, and data scientists to draft hypotheses and coordinate cross-team collaborations at scale.

3. Level up data literacy with specific curricula for personas or role families

Not every employee can be expected to be a data scientist, nor regular employees to be data literate by default. However, as firms seek to enhance the use of data for better decision-making, they should identify those roles or categories of employees that need to be comfortable with how to use and interpret data. Whether to grasp the firm's strategic context for data, to use a self-service tool, or to understand a dashboard, data literacy implies a continuous learning journey integrated into employees' career paths.

To broach this need for (re)education and (re)skilling, companies can establish training programs to reflect the specific needs of different archetypes or personas. This can be done by combining short foundational courses—like boot camps or sprints—with situated learning on data and its applications in the employee's specific business context.

Airbnb has created Data University, which trains all employees through three personas—data specialists, managers, and newbies—and across three levels of proficiency: data awareness, data collection and visualization, and data at scale. The content is geared to the specific data sets and needs of a stand-alone business unit. Uber also demonstrated a successful approach by establishing a peer-to-peer training program in which its data scientists act as coaches in various teams and work jointly with teams on specific data sets and data-heavy projects.

Incumbents typically offer trainings, but these have tended to be optional and focus mainly on data literacy and data contents outside of the context of an employee's actual business unit and functional context. Learning from the digital natives, we can see that when learning and training is tailored to personas or role families, comprising foundational as well as situated learning, it leads to more successful outcomes.

4. Advance data practices by creating communities

Data practices are typically underdeveloped at most companies, and employees assuming data roles often feel isolated. A lack of knowledge-sharing and collaboration mechanisms exacerbate their isolation.

Creating data communities around a shared domain of interest—for instance, a specific analytical tool, method, or business problem—can help promote learning and shape data practices through shared experiences. The above-mentioned self-service environments and data platforms (that is, a data library) can be used to provide advanced collaboration and support features (for example, tags, comments, documentation, contacts, requests for help).

At Netflix, product teams and data science experts discuss experimental designs (for example, A/B testing on an app interface) and their outcomes through established forums. Airbnb has a community that brings together distributed pods of data engineers through forums and working groups to support an overarching data quality initiative.

Companies need to prioritize, establish, and fund these communities to ensure data experts can engage with each other on data practices and feel less isolated. The digital natives have demonstrated that even more casual communities can gather members around a shared domain of interest (for example, reporting, analytic tools) and cause collective empowerment through experience exchange. They are also an effective way to address common issues such as data access or quality and contribute to harmonizing enterprisewide practices.

5. Promote data through various corporate communication channels

Despite some companies having success with training and educational initiatives, employee confidence in using data and instilling data into company culture are recurring barriers toward becoming a data-driven company. Educational initiatives are a precursor but are not enough on their own to change company culture on a deeper level. To increase employee confidence and stimulate demand for data, firms need to promote it.

Through a two-pronged approach, companies can raise awareness about the value and use of data. First, dedicated communication channels promote the use of data to fulfill strategic needs. Corporate events, newsletters, and podcasts are examples of channels that can also be used to advertise any new initiatives, such as training availability, successes, and so forth. Second, as companies grow their network of data users, ambassadors can be enlisted to advocate for data locally. For example, data managers can share key messages about data quality with data scientists embedded within product teams.

At Uber, we saw that leadership seeks to empower data champions as ambassadors in every business team to locally promote data value while ensuring constant communication with data experts. Netflix has created a culture charter that mentions and promotes the use of data for decision-making as a key enabler for a successful career; the organization declares data to be a part of its DNA in writing. Salesforce continuously measures the use of data insights and publishes the successes achieved.

Incumbents often haven't yet prioritized investment for promoting data and raising awareness by all channels necessary. To push data awareness, literacy, and use of data for decision-making to all corners of the organization, senior leaders and managers need to use every available communication channel or build new ones; otherwise, a data-driven culture won't translate across the organization.

As data becomes omnipresent, managers in all business functions are called on to leverage data for various applications and goals—from traditional reporting to self-service analytics to AI-driven innovation. From our regular interactions with data management executives and analytics experts, we learned that while many are aware of the challenges of building a data democracy, they are often still approaching it in a piecemeal and ad hoc manner. Building a data democracy involves coordinated orchestration and buy-in from senior leaders and managers. While these five pillars don't need to be built sequentially, they provide a foundation for moving forward. Addressing them all intentionally can help managers to successfully integrate data into all corners of the organization.

Adapted from "5 Pillars for Democratizing Data at Your Organization" on hbr.org, November 24, 2023. Reprint H07WES

8

The Age of Continuous Connection

by Nicolaj Siggelkow and Christian Terwiesch

A seismic shift is under way. Thanks to new technologies that enable frequent, low-friction, customized digital interactions, companies today are building much deeper ties with customers than ever before. Instead of waiting for customers to come to them, firms are addressing customers' needs the moment they arise—and sometimes even earlier. It's a win-win: Through what we call *connected strategies,* customers get a dramatically improved experience, and companies boost operational efficiencies and lower costs.

Consider the MagicBands that Disney World issues all its guests. These small wristbands, which incorporate radio-frequency-identification technology, allow visitors to enter the park, get priority access to rides, pay for food and merchandise, and unlock their hotel rooms. But the bands also help Disney locate guests anywhere in the park and then create customized experiences for them. Actors playing Disney characters, for example, can personally greet guests passing by ("Hey, Sophia! Happy seventh birthday!"). Disney can encourage people to visit attractions with idle capacity ("Short lines at Space Mountain right now!"). Cameras on various rides can automatically take photographs of guests, which Disney can use to create personalized memory books for them, without their ever having to pose for a picture.

Similarly, instead of just selling textbooks, McGraw-Hill Education now offers customized learning experiences. As students use the company's electronic texts to read and do assignments, digital technologies track their progress and feed data to their teachers and to the company. If someone is struggling with an assignment, her teacher will find out right away, and McGraw-Hill will direct the student to a chapter or video offering helpful explanations. Nike, too, has gotten into the game. It can now connect with customers daily, through a wellness system that includes chips embedded in shoes, software that analyzes workouts, and a social network that provides advice and support. That new model has allowed the company to transform itself from a maker of athletic gear into a purveyor of health, fitness, and coaching services.

It's easy to see how Disney, McGraw-Hill, and Nike have used approaches like these to stay ahead of the competition. Many other companies are taking steps to develop their own connected strategies by investing substantially in data gathering and analytics. That's great, but a lot of them are now awash in so much data that they're overwhelmed and struggling to cope. How can managers think clearly and systematically about what to do next? What are the best ways to use all this new information to better connect with customers?

Idea in Brief

The Old Approach

Companies used to interact with customers only episodically, when customers came to them.

The New Approach

Today, thanks to new technologies, companies can address customers' needs the moment they arise—and sometimes even earlier. With connected strategies, firms can build deeper ties with customers and dramatically improve their experiences.

The Upshot

Companies need to make continuous connection a fundamental part of their business models. They can do so with four strategies: respond to desire, curated offering, coach behavior, and automatic execution.

In our research we've identified four effective connected strategies, each of which moves beyond traditional modes of customer interaction and represents a fundamentally new business model. We call them *respond to desire, curated offering, coach behavior,* and *automatic execution.* What's innovative here is not the technologies these strategies incorporate but the ways that companies deploy those technologies to develop continuous relationships with customers.

Below, we'll define these new connected strategies and explore how you can make the most of the ones you choose to adopt. But first let's take stock of the old model they're leaving behind.

Buy What We Have

Most companies still interact with customers only episodically, after customers identify their needs and seek out products or services to meet them. You might call this model *buy what we have.* In it companies work hard to provide high-quality offerings at a competitive price and base their marketing and operations on the assumption that they'll engage only fleetingly with their customers.

Here's a typical buy-what-we-have experience: One Tuesday, working from home, David is halfway through printing a batch of urgent letters when his toner cartridge runs out. It's maddening. He *really* doesn't have time for this. Grumbling, he hunts around for his keys, gets into his car, and drives 15 minutes to the nearest office supply store. There he wanders the aisles looking for the toner section, which turns out to be an entire wall of identical-looking cartridges. After scanning the options and hoping that he recalls his printer model correctly, he finds the cartridge he needs, but only in a multipack, which is expensive. He sets off in search of a staff member who might know if the store has any single cartridges, and eventually he locates a manager, who disappears into the back of the store to check.

Much time passes. When the manager at last returns, it's to report regretfully that the store is sold out of single cartridges. Because he has to get his letters done, David decides to buy the multipack. He grabs one and heads to the checkout counter to pay, only to find himself waiting in a long line. When he finally gets home, an hour or two later, he's not a happy guy.

We find it helpful to break the traditional customer journey into three distinct stages: *recognize,* when the customer becomes aware of a need; *request,* when he or she identifies a product or service that would satisfy this need and turns to a company to meet it; and *respond,* when the customer experiences how the company delivers the product or service. At each of these stages, David suffered a lot of discomfort, but at no point along the way did the toner company have any way of learning about his discomfort or alleviating it. Company and customer were poorly connected throughout, and both parties suffered.

It doesn't have to play out that way. Each of our four connected strategies could have helped improve David's customer experience at one or more of the stages and helped the company strengthen its business.

Let's explore specifically what each strategy entails.

Respond to Desire

This strategy involves providing customers with services and products they've requested—and doing so as quickly and seamlessly as possible. The essential capabilities here are operational: fast delivery, minimal friction, flexibility, and precise execution. Customers who enjoy being in the driver's seat tend to like this strategy.

To provide a good respond-to-desire experience, companies need to listen carefully to what customers want and make the buying process easy. In many cases, what matters most to customers is the amount of energy they have to expend—the less, the better!

That's certainly what David wanted in his search for a toner cartridge. So let's imagine a respond-to-desire strategy that might serve him well in the future.

Say that upon realizing that he needs a replacement, David goes online to his favorite retailer, types in his printer model, and with just a click or two makes a same-day order for the correct cartridge. His credit card number and address are already stored in the system, so the whole process takes just a minute or two. A few hours later his doorbell rings, and he has exactly what he needs.

Speed is critical in a lot of respond-to-desire situations. Users of Lyft and Uber want cars to arrive promptly. Health care patients want the ability to connect at any time of day or night with their providers. Retail customers want the products they order online to arrive as quickly as possible—a desire that Amazon has famously focused on satisfying, in the process redefining how it interacts with customers. Years ago it set up a "one click" process for ordering and payment, and more recently it has gone even further than that. Today you can give Alexa a command to order a particular product, and she'll take care of the rest of the customer journey for you. That's responding to desire.

Curated Offering

With this strategy, companies get actively involved in helping customers at an earlier stage of the customer journey: after the customers have figured out what they need but before they've decided how to fill that need. Executed properly, a curated-offering strategy not only delights customers but also generates efficiency benefits for companies, by steering customers toward products and services that firms can easily provide at the time. The key capability here is a personalized recommendation process. Customers who value advice—but still want to make the final decision—like this approach.

How might a curated-offering strategy serve David? Consider this scenario: He goes online to order his toner cartridge, and the site automatically suggests the correct one on the basis of what he has bought before. That spares him the hassle of finding the model number of his printer and figuring out which cartridge he needs. So now he just orders what the site suggests, and a few hours later, when his doorbell rings, he's had his needs smoothly and easily met.

Blue Apron and similar meal-kit providers have very effectively adopted the curated-offering strategy. This differentiates them from Instacart and many of the other grocery delivery services that have emerged in recent years, all of which are guided by a "you order, we deliver" principle—in other words, a respond-to-desire strategy. The Instacart approach might suit you better than spending time in a supermarket checkout line, but it doesn't relieve you of the burden of hunting for recipes and creating shopping lists of ingredients. Nor does it prevent you from overbuying when you do your shopping. Blue Apron helps on all those fronts, by presenting you with personally tailored offerings, creating an experience that many people find is more convenient, fun, and healthful than what they would choose on their own.

Coach Behavior

Both of the previous two strategies require customers to identify their needs in a timely manner, which (being human) we're not always good at. Coach-behavior strategies help with this challenge, by proactively reminding customers of their needs and encouraging them to take steps to achieve their goals.

Coaching behavior works best with customers who know they need nudging. Some people want to get in shape but can't stick to a workout regimen. Others need to take medications but are forgetful. In these situations a company can watch over customers and help them. Knowledge of a customer's needs might come from information that the person has previously shared with the firm or from observing the behavior of many customers. The essential capabilities involved are a deep understanding of customer needs ("What does the customer really want to achieve?") and the ability to gather and interpret rich contextual data ("What has the customer done or not done up to this point? Can she now enact behaviors that will get her closer to her goal?").

Here's what a coach-behavior strategy for David might look like: Perhaps the printer itself tracks the number of pages it has generated since David last changed the toner and sends that information back to the manufacturer, which knows that he will soon need a new cartridge. So it might email him a reminder to reorder. At the same time, it might encourage him to run the cleaning function on his printer—a suggestion that will help him avoid later inconveniences. Coached in this way, David will have his new printer cartridge before the old one runs out; he'll lose almost no time in replacing it; and he'll have a clean printer that performs at its best.

To implement coach-behavior approaches well, a company needs to receive information constantly from its customers so that it doesn't miss the right moment to suggest action. The technical challenge in this sort of relationship lies in enabling cheap and reliable two-way communication with customers. Traditionally, this had been difficult, but it's getting easier all the time. The advent of wearable devices, for example, allows health care companies to hover digitally over customers around the clock, constantly monitoring how they're doing.

Nike's new business model incorporates coach-behavior strategies. By making its customers part of virtual running clubs and tracking their runs, the company knows when it's time for their next workout, and through its app it can offer them audio training guides and plans. This kind of timely and personal connection builds trust and encourages customers to think of Nike as a health-and-fitness coach rather than just a shoe manufacturer, which in turn means that when the company's app nudges them to run, they're more likely to do it. This serves customers well, because it keeps them motivated and in shape. And it serves Nike well, of course, because customers who run more buy more shoes.

Automatic Execution

All the strategies we've discussed so far require customer involvement. But this last strategy allows companies to meet the needs of customers even before they've become aware of those needs.

In an automatic-execution strategy, customers authorize a company to take care of something, and from that point on the company handles everything. The essential elements here are strong trust, a rich flow of information from the customers, and the ability to use it to flawlessly anticipate what they want. The customers most open to automatic execution are comfortable having data stream constantly from their devices to companies they buy from and have faith that those companies will use their data to fulfill their needs at a reasonable price and without compromising their privacy.

Here's how automatic execution might work for David. When he buys his printer, he authorizes the manufacturer to remotely monitor his ink level and send him new toner cartridges whenever it gets low. From then on, the onus is on the company to manage his needs, and David is spared several hassles: recognizing that he's low on toner, figuring out how to get more, and buying it. Instead, he just goes about his business. When the time is right, his doorbell will ring, and he'll have exactly what he needs.

The growing internet of things is making all sorts of automatic execution possible. David's printer cartridge scenario isn't just hypothetical: Both HP and Brother already have programs that ship replacement toner to customers whenever their printers send out a "low ink" signal. Soon our refrigerators, sensing that we're almost out of milk, will be able to order more for delivery by tomorrow morning—but naturally only after checking our calendar to make sure we're not going on a vacation and wouldn't need milk after all.

Automatic execution will make people's lives easier and in some cases will even save lives. Consider fall-detection sensors, the small medical devices worn by many seniors. Initially, the companies who made them did so using the respond-to-desire model. If an elderly person who was wearing one fell and needed help, she could press a button that activated a distress call. That was good, but it didn't work if someone was too incapacitated to press the button. Now, though, internet-connected wearable technologies allow health care companies to monitor patients constantly in real time, which means people don't need to actively request assistance if and when they're in distress. Imagine a bracelet that monitors vital signs and uses an accelerometer to detect falls. If a person wearing the bracelet slips, tumbles down the basement stairs, and is knocked unconscious, the bracelet's sensor will immediately detect the emergency and summon help. That's automatic execution.

We're excited about automatic execution, but we want to stress that we don't see it as the best solution to all problems—or

Which connected strategies should you use?

Connected strategy	Description	Key capability	Works best when	Works best for
Respond to desire	Customer expresses what she wants and when	Fast and efficient response to orders	Customers are knowledgeable	Customers who don't want to share too much data and who like to be in control
Curated offering	Firm offers tailored menu of options to customer	Making good personalized recommendations	The uncurated set of options is large and potentially overwhelming	Customers who don't mind sharing some data but want a final say
Coach behavior	Firm nudges customer to act to obtain a goal	Understanding customer needs and ability to gather and interpret rich data	Inertia and biases keep customers from achieving what's best for them	Customers who don't mind sharing personal data and getting suggestions
Automatic execution	Firm fills customer's need without being asked	Monitoring customers and translating incoming data into action	Customer behavior is very predictable and costs of mistakes are small	Customers who don't mind sharing personal data and having firms make decisions for them

for all customers. People differ in the degree to which they feel comfortable sharing data and in having the companies serving them act on that data. One family might be delighted to receive an automatically generated personal memory book after a visit to Disney World, but another might think it's creepy and invasive. If companies want customers to make a lot of personal data available on an automated and continuous basis, they will need to prove themselves worthy of their customers' trust. They'll need to show customers that they'll safeguard the privacy and security of personal information and that they'll only recommend products and services in good faith. Breaking a customer's trust at this level could mean losing that customer—and possibly many other customers—forever.

A final important point: Given that companies are likely to have customers with different preferences, most firms will have to create a portfolio of connected strategies, which will require them to build a whole new set of capabilities. (See the table "Which connected strategies should you use?") One-size-fits-all usually won't work.

Repeat

Earlier, we mentioned that we like to think of the individual customer journey as having three stages: *recognize, request,* and *respond.* But there's actually a fourth stage—*repeat*—which is fundamental to any connected strategy, because it transforms stand-alone experiences into long-lasting, valuable relationships. It is in this stage that companies learn from existing interactions and shape future ones—and discover how to create a sustainable competitive advantage.

The repeat dimension of a connected strategy helps companies with two forms of learning.

First, it allows a company to get better at matching the needs of an individual customer with the company's existing products and services. Over time and through multiple interactions, Disney sees that a customer seems to like ice cream more than fries, and theater performances more than fast rides—information that then allows the company to create a more enjoyable itinerary for him. McGraw-Hill sees that a student struggles with compound-interest calculations, which lets it direct her attention to material that covers exactly that weakness. Netflix sees that a customer likes political satire, which allows it to make pertinent movie suggestions to her.

Second, in the repeat stage companies can learn at the population level, which helps them make smart adjustments to their portfolios of products and services. If Disney sees that the general demand for frozen yogurt is rising, it can increase the number of stands in its parks that serve frozen yogurt. If McGraw-Hill sees that many students are struggling with compound-interest calculations, it can refine its online module on that topic. If Netflix observes that many viewers like political dramas, it can license or produce new series in that genre.

Both of these loops have positive feedback effects. The better the company understands a customer, the more it can customize its offerings to her. The more delighted she is by this, the more likely she is to return to the company again, thus providing it with even more data. The more data the company has, the better it can customize its offerings. Likewise, the more new customers a company attracts through its superior customization, the better its population-level data is. The better its population data, the more it can create desirable products. The more desirable its products, the more it can attract new customers. And so on. Both learning loops build on themselves, allowing companies to keep expanding their competitive advantage.

Over time these two loops have another very important effect: They allow companies to address more-fundamental customer needs and desires. McGraw-Hill might find out that a customer wants not just to understand financial accounting but also to have a career on Wall Street. Nike might find out that a particular runner is interested not just in keeping fit but also in training to run a first marathon. That knowledge offers opportunities for companies to create an even wider range of services and to develop trusted relationships with customers that become very hard for competitors to disrupt.

We can't tell you where all this is headed, of course. But here's what we know: The age of "buy what we have" is over. If you want to achieve sustainable competitive advantage in the years ahead, connected strategies need to be a fundamental part of your business. This holds true whether you're a startup trying to break into an existing industry or an incumbent firm trying to defend your market, and whether you deal directly with consumers or operate in a business-to-business setting. The time to think about connected strategies is now, before others in your industry beat you to it.

Originally published in May–June 2019. Reprint R1903C

9

Want Your Company to Get Better at Experimentation?

by Iavor Bojinov, David Holtz, Ramesh Johari, Sven Schmit, and Martin Tingley

F or years online experimentation has fueled the innovations of leading tech companies such as Amazon, Alphabet, Meta, Microsoft, and Netflix, enabling them to rapidly test and refine new ideas, optimize product features, personalize user experiences, and maintain a competitive edge. Owing to the widespread availability and lower cost of experimentation tools today, most organizations—even those outside the technology sector—conduct online experiments.

However, many companies use online experimentation for just a handful of carefully selected projects. That's because their data scientists are the only ones who can design, run, and analyze tests. It's impossible to scale up that approach, and scaling matters. Research from Microsoft (replicated at other companies) reveals that teams and companies that run lots of tests outperform those that conduct just a few, for two reasons: Because most ideas have no positive impact, and it's hard to predict which will succeed, companies must run lots of tests. And as the growth of AI—particularly generative AI—makes it cheaper and easier to create numerous digital product experiences, they must vastly increase the number of experiments they conduct—to hundreds or even thousands—to stay competitive.

Scaling up experimentation entails moving away from a data-scientist-centric approach to one that empowers *everyone* on product, marketing, engineering, and operations teams—product managers, software engineers, designers, marketing managers, and search-engine-optimization specialists—to run experiments. But that presents a challenge. Drawing on our experience working for and consulting with leading organizations such as Airbnb, LinkedIn, Eppo, Netflix, and Optimizely, we provide a road map for using experimentation to increase a company's competitive edge by (1) transitioning to a self-service model that enables the testing of hundreds or even thousands of ideas a year and (2) focusing on hypothesis-driven innovation by both learning from individual experiments and learning *across* experiments to drive strategic choices on the basis of customer feedback. These two steps in tandem can prepare organizations to succeed in the age of AI by innovating and learning faster than their competitors do. (The opinions expressed in this article are ours and do not represent those of the companies we have mentioned.)

The Current State

The basics of experimentation are straightforward. Running an A/B test involves three main steps: creating a challenger (or variant) that deviates from the status quo; defining a target population (the subset of customers targeted for the test); and selecting a

Idea in Brief

The Problem

Many companies struggle to make experimentation a consistent, scalable part of how they operate. Efforts often remain isolated to innovation labs or digital teams, limiting their impact on core business decisions.

The Solution

To embed experimentation into the fabric of the organization, executives must build systems that support it at scale. This includes creating the right culture, incentives, and infrastructure to support experiments, as well as protecting them from political or performance pressures. Leaders must also establish strong governance to ensure experiments are aligned with strategic goals.

The Payoff

When experimentation becomes a disciplined, organizationwide capability, companies can make smarter decisions, adapt faster, and unlock continuous innovation.

metric (such as product engagement or conversion rate) that will be used to assess the outcome. Here's an example: In late 2019, when one of us (Martin) led its experimentation platform team, Netflix tested whether adding a Top 10 row (the challenger) on its user interface to show members (the target population) the most popular films and TV shows in their country would improve the user experience as measured by viewing engagement on Netflix (the outcome metric). The experiment revealed that the change did indeed improve the user experience without impairing other important business outcomes, such as the number of customer service tickets or user-interface load times. So the Top 10 row was released to all users in early 2020. As this example illustrates, experimentation enables organizations to make data-driven decisions on the basis of observed customer behavior.

Barriers to Scaling Up Experimentation

Data science teams often lead the adoption of online experimentation. After initial success, organizations tend to fall into a rut, and the returns remain limited. A common pattern we see is this: The organization invests in a platform technically capable of designing, running, and analyzing experiments. Large technology companies build their own platforms in-house; others typically buy them from vendors. Although these tools are widely available, investing in them is costly. Building a platform can take more than a year and usually requires a team of five to 10 engineers. External platforms generally cost less and are faster to implement, but they still require dedicated resources to be integrated with the organization's internal development processes and to gain approval from legal, finance, and cybersecurity departments.

After the initial investment, leaders who sponsored the platform (usually the heads of data science and product) face pressure to quickly demonstrate its value by scoring successes—experiments that yield statistically significant positive results in favor of the challenger. In an attempt to avoid negative results, they try to anticipate which ideas will have a big impact—something that is exceptionally difficult to predict. For example, in late 2012, when Airbnb launched its neighborhood travel guides (web pages listing things to do, best restaurants, and so on), the content was heavily viewed, but overall bookings declined. In contrast, when the company introduced a trivial modification—the ability to open an accommodation listing in a new browser tab rather than the existing one, which made it easier to compare multiple listings—bookings increased by 3% to 4%, making it one of the company's most successful experiments.

Motivated to turn every experiment into a success, teams often overanalyze each one, with data scientists spending more than 10 hours per experiment. The results are disseminated in memos and discussed in product-development meetings, consuming many hours of employee time. Although the memos are broadly available in principle, the findings they contain are never synthesized to identify patterns and generalizable lessons; nor are they archived in a standardized fashion. As a result, it's not uncommon for different teams (or even the same team after its members have turned over) to repeatedly test an unsuccessful idea.

Looking to increase the adoption of and returns from experimentation, data science and product leaders tend to focus on incremental changes: increasing the size of product teams so as to run more experiments and more easily prioritize which ideas to test; hiring additional data scientists to analyze the increased number of tests and reduce the time needed to execute on them; and instigating more knowledge-sharing meetings for the dissemination of results. In our experience, however, those tactics are unsuccessful. Managers struggle to identify which tests will lead to a meaningful impact; hiring more data scientists provides only a marginal increase in experimentation capacity; and knowledge-sharing meetings don't create institutional knowledge. These tactics may appear sensible, but they end up limiting the adoption of experimentation because the processes they establish don't scale up.

Democratizing Experimentation

To achieve enterprise-wide experimentation for data-driven decisions, companies have to transition to a self-service approach: empower all employees on the product, marketing, engineering, and operations teams to test changes small and large and then learn from and act on outcomes. That means embedding some important functions in the platform and redesigning the data scientists' jobs.

The platform. The data science organization (data scientists, data engineers, and software engineers) should ensure that the platform contains the following features, whether it is built internally or purchased.

- *A simple, easy-to-understand interface.* Airbnb had such a system, which enabled a single engineer to implement and test the feature that opened accommodation listings in a new tab.

- *The ability to automatically impose statistical rigor.* Tasks such as determining the appropriate duration for a particular type of experiment and the criteria for deciding whether the results are significant should be automated using historical data.

- *Embedded experimentation protocols.* Instructions should provide default settings for most aspects of standard experiments, such as decision-metric selection. These protocols allow users to design and launch experiments with minimal input from data scientists.

- *Automated rollbacks.* These are quantitative criteria that act as trip wires to stop an experiment if its impact is too negative—for example, a significant drop in the number of daily active users of a social media site. The impact is measured using guardrail metrics—secondary measurements that ensure that while you're focused on improving one outcome, you don't unintentionally harm other important areas such as user experience, revenue, or system stability. When a vast number of experiments are running concurrently, such a feature is vital.

- *An AI assistant that provides easy-to-understand explanations of complex concepts.* This core element can simplify the design and analysis of experiments, making the process accessible even to novice users.

Data scientists' role. In addition to setting up the platform, data scientists should be responsible for training employees, creating the materials for that training, and holding office hours to answer complex questions after everyone is up and running. The time they spend on most tests will drop to nearly zero because they will no longer be involved in execution or analysis. (They will still be involved in novel tests, such as the first in a new product space, and will be called in when results are challenging to interpret. But those are the exceptions.) Thus they can focus on projects of greater impact that leverage their unique expertise: for example, developing new statistical methods for analyzing complex experiments and analyzing company data in light of past test results to identify new possibilities for product initiatives.

Preparing the Organization

In organizations that have not adopted experimentation, product teams are generally evaluated according to whether they launch new products. When they start experimenting, too often the criterion becomes the number of "successful" experiments run. Unfortunately, that makes employees risk-averse, so they run too few experiments. Scaling up experimentation, therefore, requires changing incentives. Companies should evaluate employees on the basis of the overall performance of the business unit and the organization, not the outcome of individual tests.

That shift will encourage a far wider range of employees to generate and test as many ideas as possible, increasing their chances of discovering breakthroughs that enhance performance. But it will also result in testing potentially higher-risk ideas with less oversight from experienced data scientists—something that can make people hesitant to run experiments. As we mentioned, one solution is to embed guardrails (quantitative criteria that act as trip wires) in the platform. Another is to roll out new features or changes in phases—a practice common among the largest tech firms. For example, updates to mobile apps from the Apple App Store and the Google Play Store are released that way to reduce risk.

Hypothesis-Driven Innovation

As organizations adopt and scale up experimentation throughout the enterprise and transition to an incentive model that rewards overall business impact, product leaders should be able to extract significantly more value by focusing on understanding the *why* behind test results. That requires managers to use experimentation for more than making data-driven decisions—such as whether a particular change is better than the status quo—by hypothesizing *why* that is so. The experiment allows them to test the theory; by considering additional metrics, they can understand the mechanism that drove the result. Crucially, a focus on *why* fuels more customer-centric innovation, because feedback—gathered through experiments—is consulted not only to choose between the variant and the status quo but also to determine the next experiment and the overall product direction.

Netflix's Top 10 experiment, for instance, began with a clear hypothesis: The Top 10 row would help members find content to watch by tapping into an innate desire for shared experiences and conversations. That would increase member joy and satisfaction, as measured by increased member engagement. In addition to tracking overall engagement, the experiment monitored metrics such as where members found content (Search, My List, various rows on the home page) and how they interacted with the titles showcased in the Top 10 row. (Those titles were also available in the status quo experience but in a different location.) The additional metrics demonstrated how members changed their behavior in response to the new row. For example, because Netflix aims to connect members with the best content for them directly from the home page, an increased use of Search in response to the Top 10 row would indicate that the home page had not been delivering on that goal. That information would be used to design a subsequent test.

Once an organization is running hundreds or thousands of experiments a year, however, it becomes impossible to review every one of them in dedicated memos and meetings. Organizations should therefore shift their focus from analyzing individual experiments to analyzing, discussing, and learning from groups of related experiments, such as those concerning the search function or product-details pages that provide pictures, specifications, reviews, and other information. We refer to such efforts as *experimentation programs*. This shift is the key to unlocking significant additional value from experimentation.

When experiments are considered in this way, an organization can embrace more-efficient, hypothesis-driven innovation practices that build on prior tests to inform future ones. Experimentation programs also encourage product teams to break complex ideas down into small, testable hypotheses, making it easier to adapt the direction of a product to customer demands.

Experimentation Programs

Once an organization has become competent at learning across experiments, the next step is to compare results across experimentation programs, which makes it possible to evaluate the relative performance of various product areas and identify potential investment opportunities. Consider an e-commerce platform that has multiple features designed to help shoppers find the right product, two of which are the search function and the product-details page. The business would most likely have one experimentation program for search and another for product pages.

Now suppose that changes to the ranking algorithm used in a search engine generated positive but diminishing returns, as measured by the effect on sales in successive experiments. Meanwhile, all but one of the tests on the product-details page consistently showed small negative effects on sales, and that one exception produced large positive effects. One big "win" for the product-details page amid a number of unsuccessful tests suggests that the company doesn't yet understand what aspects of product description resonate most with customers. Additional resources should be devoted to that experimentation program. Meanwhile, the diminishing returns on search-ranking experiments suggest a mature search-engine algorithm; leaders should consider either exploring vastly different approaches—such as an AI chatbot—or shifting resources to other areas for experimentation, such as product-details pages.

A Knowledge Repository

Learning across experiments at scale requires creating a knowledge repository—a system designed to store, categorize, and organize experiment results (including effects on sales and other key metrics, hypotheses about impacts on customers, and so on)—and making the information in it accessible to data scientists, product managers, and leadership. A repository allows the organization not only to track the state of any experimentation program but also to spread learning across the enterprise, which is crucial for hypothesis-driven innovation when a company is running a huge number of experiments each year.

A knowledge repository should perform four key functions: (1) It should make it possible to group experiments into programs. Many organizations would most likely group them by feature (such as search engine or product details) or business unit (such as marketing or customer support). (2) It should store and track the KPIs (quantity sold, revenue, conversions, and so on) that are important across the business. That will allow the impact of various experiments and experimentation programs to be compared on common terms. For example, most of Netflix's experiments are designed to improve one of a handful of KPIs, such as engagement. (3) It should host all documents related to each test, mapping them to the experimentation program to ensure that all learnings are centrally available. (4) Most important, it should enable all employees to easily extract insights. Dashboards that track the performance of experimentation programs (such as the number of experiments run, the number of feature changes rolled out to the entire user base, and the cumulative impact of experiments on users over the previous quarter) are a great starting point. However, a more dynamic access point is an "assistant" powered by generative AI that can answer complex questions about past experiments.

. . .

Leading tech organizations use experimentation to innovate and improve performance rapidly by testing all ideas—not just carefully vetted ones or only the big ones. Moreover, learnings from those experiments (often gleaned from combining results across similar experiments) generate new ideas for testing. Experimentation can be scaled up only by democratizing access to tools, aligning incentives with improvements in long-term outcomes, and enabling employees to easily view, compare, and synthesize the results of experiments both within and across experimentation programs. Thanks to modern data tools and advances in AI, becoming expert in experimentation is now within reach for many more organizations. Given that the same AI advances are reducing the cost of coming up with, testing, and building innovative product variants, leaders must turn what is possible into a reality in their organizations.

Originally published in January–February 2025. Reprint R2501G

10

Reskilling in the Age of AI

by Jorge Tamayo, Leila Doumi, Sagar Goel, Orsolya Kovács-Ondrejkovic, and Raffaella Sadun

B ack in 2019 the Organisation for Economic Co-operation and Development made a bold forecast. Within 15 to 20 years, it predicted, new automation technologies were likely to eliminate 14% of the world's jobs and radically transform another 32%. Those were sobering numbers, involving more than 1 billion people globally—and they didn't even factor in ChatGPT and the new wave of generative AI that has recently taken the market by storm.

Today, advances in technology are changing the demand for skills at an accelerated pace. New technologies can not only handle a growing number of repetitive and manual tasks but also perform increasingly sophisticated kinds of knowledge-based work—such as research, coding, and writing—that have long been considered safe from disruption. The average half-life of skills is now less than five years, and in some tech fields it's as low as two and a half years. Not all knowledge workers will lose their jobs in the years ahead, of course, but as they carry out their daily tasks, many of them may well discover that AI and other new technologies have so significantly altered the nature of what they do that in effect they're working in completely new fields.

To cope with these disruptions, a number of organizations are already investing heavily in upskilling their workforces. One recent BCG study suggests that such investments represent as much as 1.5% of those organizations' total budgets. But upskilling alone won't be enough. If the OECD estimates are correct, in the coming decades millions of workers may need to be entirely *reskilled*—a fundamental and profoundly complex societal challenge that will require workers not only to acquire new skills but to use them to change occupations.

Companies have a critical role to play in addressing this challenge, and it's in their best interests to get going on it in a serious way right now. Among those that have embraced the reskilling challenge, only a handful have done so effectively, and even *their* efforts have often been subscale and of limited impact, which leads to a question: Now that the need for a reskilling revolution is apparent, what must companies do to make it happen?

In our work at the HBS Digital Reskilling Lab and the BCG Henderson Institute we have been studying this question in depth, and as part of that effort we interviewed leaders at almost 40 organizations around the world that are investing in large-scale reskilling programs. During those interviews we discussed common challenges, heard stories of early success, and discovered that many of those companies are thinking in important new ways about why, when, and how to reskill. In synthesizing what we've learned, we've become aware of five paradigm shifts that are emerging in reskilling—shifts that companies will need

Idea in Brief

The Situation

New technologies can not only handle a growing number of repetitive and manual tasks but also perform sophisticated kinds of knowledge-based work—such as research, coding, and writing—that have long been considered safe from disruption.

The Challenge

To cope, many organizations are investing heavily in upskilling their workforces, but those efforts alone won't be enough. In the coming decades millions of workers may need to be entirely reskilled—a profoundly complex societal challenge.

The Path Forward

Some companies have recently launched successful reskilling efforts. Five important paradigm shifts have emerged from their efforts that other companies will need to embrace if they hope to adapt to the new era of automation and AI.

to understand and embrace if they hope to succeed in adapting dynamically to the rapidly evolving era of automation and AI.

In this article we'll explore those shifts. We'll show how some companies are implementing them, and we'll review the unexpected challenges they've encountered and the promising wins they've achieved.

1. Reskilling Is a Strategic Imperative

During times of disruption, when many jobs are threatened, companies have often turned to reskilling to soften the blow of layoffs, assuage feelings of guilt about social responsibility, and create a positive PR narrative. But most of the companies we spoke with have moved beyond that narrow approach and now recognize reskilling as a strategic imperative. That shift reflects profound changes in the labor market, which is increasingly constrained by the aging of the working population, the emergence of new occupations, and an increasing need for employees to develop skills that are company-specific. Against this backdrop effective reskilling initiatives are critical, because they allow companies to build competitive advantage quickly by developing talent that is not readily available in the market and filling skills gaps that are instrumental to achieving their strategic objectives—before and better than their competitors do.

In recent years several major companies have embraced this approach. Infosys, for example, has reskilled more than 2,000 cybersecurity experts with various adjacent competencies and capability levels. Vodafone aims to draw from internal talent to fill 40% of its software developer needs. And Amazon, through its Machine Learning University, has enabled thousands of employees who initially had little experience in machine learning to become experts in the field.

Some companies now consider reskilling a core part of their employee value proposition and a strategic means of balancing workforce supply and demand. At those companies employees are encouraged to reskill for roles that appeal to them. Mahindra & Mahindra, Wipro, and Ericsson have policies, tools, and IT platforms that promote reskilling resources and available jobs—as does McDonald's, where restaurant employees have access to an app called Archways to Opportunity that maps skills learned on the job to career paths within the company and in other industries.

Finally, some companies are using reskilling to tap into broader talent pools and attract candidates who wouldn't otherwise be considered for open positions. ICICI Bank—headquartered in

Mumbai and employing more than 130,000 people—runs an intense, academy-like reskilling program that prepares graduates, often from diverse backgrounds, for frontline managerial jobs. The program reskills some 2,500 to 4,000 employees each year. CVS used a similar approach during the Covid-19 pandemic to hire, train, and onboard people (some of them laid-off hospitality workers) to create capacity for its critical vaccine and testing services.

2. Reskilling Is the Responsibility of Every Leader and Manager

Traditionally, reskilling is considered part of the overall corporate-learning function. When that's the case, responsibility for the design and implementation of the program is often siloed within HR, and its failure or success is measured very narrowly—in terms of the number of trainings delivered, the cost per learner, and similar training-specific metrics. According to a recent BCG report, only 24% of polled companies make a clear connection between corporate strategy and reskilling efforts. Reskilling investments need a profound commitment from HR leaders, of course, but unless the rest of the organization understands the strategic relevance of those investments, it's very hard to obtain the relentless and distributed effort that such initiatives require to succeed.

At most of the organizations where we interviewed, reskilling initiatives are visibly championed by senior leaders, often CEOs and chief operating officers. They work hard to articulate for the rest of the company the connection between reskilling and strategy and to ensure that leadership and management teams understand their shared responsibility for implementing these programs. For example, as part of its ongoing digital transformation, Ericsson has developed a multiyear strategy devoted to upskilling and reskilling. The effort involves systematically defining critical skills connected to strategy, which correspond to a variety of accelerator programs, skill journeys, and skill-shifting targets—most of them dedicated to transforming telecommunications experts into AI and data-science experts. The company considers this a high-priority, high-investment project and has made it part of the objectives and key results that executives review quarterly. In just three years Ericsson has upskilled more than 15,000 employees in AI and automation.

Similarly, the executive team at CVS has made training and reskilling an integral part of the company's business strategies. Each individual business leader is now responsible for designing and delivering workforce-reskilling plans to help the company reach its goals, and the ability to do so is factored in to performance assessments. Amazon, too, has famously committed to reskilling as a core strategic objective and now mentions it prominently in its leadership manifesto for managers. The visibility of this commitment contributes to Amazon's ability to achieve scale in reskilling programs.

3. Reskilling Is a Change-Management Initiative

To design and implement ambitious reskilling programs, companies must do a lot more than just train employees: They must create an organizational context conducive to success. To do that they need to ensure the right mindset and behaviors among employees and managers alike. From this perspective, reskilling is akin to a change-management initiative, because it requires a focus on many different tasks simultaneously.

Let's consider several of the most important.

Understanding supply and demand

To create a successful reskilling program, companies need a sophisticated understanding of supply (skills available internally and externally) and demand (skills needed to beat the competition). A useful way to develop this understanding is with a "skill taxonomy"—a detailed description of the capabilities needed for each occupation at a company. Employers used to put a lot of effort into creating such taxonomies from scratch, but many leading companies now rely on external providers for the bulk of the work. HSBC, for example, has adopted the taxonomy published by the World Economic Forum and customized it slightly to add skills specific to parts of its business. Similarly, SAP, which used to maintain an in-house taxonomy of 7,000 skills, has recently started working with Lightcast, which keeps a continually updated skill database. But developing a skill taxonomy is only the first step. Next comes the difficult job of deciding which skills get mapped to which jobs. Managers from different divisions may disagree about this. Such disagreement is often symptomatic of a deeper misalignment, and companies will need to resolve that before they undertake any major reskilling initiative.

Leaders must also determine what skills they will need in the future—a dynamic process that's critical for strategic reskilling programs. To do that well, they should focus on figuring out what skills the current strategy demands. Here they'll need to develop a rigorous strategic workforce-planning methodology. The European insurance company Allianz has done interesting work on this front: It regularly translates forecasted business growth into talent demand, focusing on the number of people needed in various jobs and the skills they'll require. The model, which is updated as part of the annual planning process, involves economic scenario planning and takes into account the possible effects of digitalization on the workforce.

Recruiting and evaluating

Traditionally, candidates are recruited for training opportunities or internal roles on the basis of their degrees or relevant work experience, but that obviously doesn't work for reskilled workers. A well-developed skill taxonomy can help here, by allowing organizations to think about enrollment policies in light of skill adjacencies, which can facilitate the transition from one skill set to another. Novartis has implemented an AI-powered internal talent marketplace that predicts, matches, and offers roles and projects related to employees' skills and goals. In our research we've also found that if reskilling programs are to succeed, companies must develop a clear set of enrollment criteria for employees, not all of whom will have the right combination of motivation and personality traits to be a good fit for reskilling.

Shaping the mindset of middle managers

Middle managers are often resistant to the idea of reskilling, for two main reasons: They worry (1) that their reports won't be able to keep up with their regular responsibilities while being reskilled, and (2) that once their reports *are* reskilled, they'll move to other parts of the organization. In both cases this can lead to "talent hoarding," in which managers try to hold on to their favorite reports by denying them the ability to participate in reskilling. Several of the companies we spoke with have addressed this problem by making talent development an explicit managerial responsibility. Wipro evaluates managers according to their teams' participation in training offers, and Amazon promotes leaders on the basis of a performance assessment that includes the question "How have you developed your team?" Middle managers may also resist the idea of hiring reskilled employees, believing that they're not as desirable as traditionally skilled workers. This problem can be addressed by involving managers in the design and delivery of reskilling programs and by providing sensitivity and unconscious-bias training. No matter what form the resistance takes, senior leaders' role modeling in support of reskilling is vital to overcoming it.

Building skills in the flow of work

It can be costly and logistically challenging to take employees away from their day jobs to participate in training. And adults tend not to like or learn well in classroom-style situations. In a 2021 BCG survey 65% of the 209,000 participating workers said they prefer to learn on the job. As a result, the best approach for reskilling is to do as much training as possible by means of shadowing assignments, internal apprenticeships, and trial periods. The reskilling program at ICICI Bank, for example, consists of a four-month vocational residency, during which employees take part in simulation-style trainings for the managerial role they hope to get, and an eight-month deployment in the field that involves a structured internship in a bank branch and closely shadowing a current manager.

Matching and integrating reskilled employees

Employees need to be matched with new jobs. Our interview data shows that if destination roles are clearly described in advance, employees become more interested in reskilling because new career trajectories become apparent to them, and the reskilling itself becomes more effective because it's more position-specific. Once in their new jobs, reskilled employees need several kinds of support to integrate successfully: help with learning new work norms and culture, building networks, and developing soft skills. Here coaching and mentoring can be particularly effective tools. Amazon has demonstrated leadership in this area: It runs a variety of mentoring programs for reskilled employees, among them a buddy system, part of its Grow Our Own Talent program, that connects previous and current program participants. The company also provides career coaching for employees who are making particularly difficult transitions, such as from warehouse worker to software developer.

4. Employees *Want* to Reskill—When It Makes Sense

Many of the companies we spoke with mentioned that one of their biggest challenges was simply persuading employees to embark on reskilling programs. That's understandable: Reskilling requires a lot of effort and can set a major life change in motion, and the outcome isn't guaranteed. The OECD reports that only a very small fraction of workers typically take part in standard training programs, and those who do are often the ones who need them the least.

But workers may be more willing to engage in reskilling than prior data suggests. BCG data shows, for example, that 68% of workers are aware of coming disruptions in their fields and are willing to reskill to remain competitively employed. The key to success in this domain, our interviews suggest, is to treat workers respectfully and make the benefits of their participation in reskilling initiatives clear. As one of our interviewees explains,

"The secret to scaling up reskilling programs is to design a product your employees actually like."

So how can organizations do that? We have several suggestions.

Treat employees as partners

Because reskilling programs are often associated with organizational disruption and job loss—or at least job change—leaders often avoid talking openly about the rationale for the programs and the opportunities they present. But employees are more likely to participate if they understand why the programs are being implemented and have had a role in creating them. Aware of this, several of the companies we spoke with made a point of being honest and clear about why they were creating reskilling programs and involving workers early. One large auto manufacturer, for example, told its diesel engineers that because of changes in the automobile industry, it had less and less need for their skills; it presented its program as a way of ensuring that they would have new jobs and job security in the years ahead. The companies also told us that in designing and implementing reskilling programs, it's critical to align with worker councils and unions early on and to involve them in advocating for the programs.

Design programs from the employee point of view

Reskilling programs require participants to make a major investment of time. So it's important to try to reduce the risk, cost, and effort involved and to provide (almost) guaranteed outcomes. Amazon allows employees in its Career Choice program to pursue everything from bachelor's degrees to certificates—and covers all costs in advance. That has proved to be a key factor in scaling up the program, which has already had more than

130,000 participants. CVS, for its part, uses an effective "train in place" model for new employees.

Dedicate adequate time and attention to the task

Because reskilling involves occupational change, it usually requires intensive learning, which is possible only if employees have the time and mental space they need to succeed. To that end, four times a year Vodafone dedicates days during which employees may devote themselves entirely to learning and personal development. Bosch goes even further: To help traditional engineers at the company earn degrees and get training in emerging fields, its Mission to Move program covers the cost of tuition and time spent learning for as much as two days a week for a whole year. It even gives participants days off before exams to prepare.

Naturally, providing employees the time and space for skilling can be harder in industries where most workers are hourly or shift-based. Iberdrola, a renewable energy company, faced this challenge as it digitized. Because it was embracing new technologies, the company realized it would need to reskill 3,300 employees in various hourly roles. Its leaders got the job done by working closely with frontline managers to ensure that operations weren't disrupted by workers taking time off for training. The company considered all training hours to be work hours and paid employees for them accordingly.

5. Reskilling Takes a Village

Companies have tended to think of reskilling as an organization-level challenge, believing that they have to do the job by and for themselves. But many of the companies where we interviewed have recognized that reskilling takes place in an ecosystem in which a number of actors have roles to play. Governments can incentivize reskilling investments by means of funds, policies, and public programs; industry can team up with academia to develop new skill-building techniques; and NGOs can play a role in connecting corporate talent needs with disadvantaged and marginalized talent groups. Coalitions of companies may be more effective at the reskilling challenge than single organizations are.

When designing reskilling programs for the rapidly evolving era of AI and automation, companies need to harness the potential of this wider ecosystem. We've identified several ways in which they can do so.

Consider industry partnerships

Instead of thinking of themselves as competitors for a limited talent pool, companies can team up to conduct joint training efforts, which may significantly attenuate some of the challenges outlined above. For example, industrywide skill taxonomies would provide a useful infrastructure and could in some cases help companies pool the knowledge and resources needed to invest in certain types of capabilities, such as cutting-edge AI skills, which are so new that individual organizations may not yet have the knowledge or the capacity to develop solutions on their own. Industry coalitions could also reassure participants that their investments in learning might open up broader future opportunities.

The Technology in Finance Immersion Programme, offered by the Institute of Banking and Finance Singapore, a nonprofit industry association, is a case in point. The program aims to build up an industry pipeline of capabilities in key technology areas, with participation from all major banks, insurance players, and asset managers in the country, to meet the talent needs of the financial services sector. Similarly, within the European Union a variety of stakeholders have formed the Automotive Skills Alliance, which is dedicated to the "re-skilling and up-skilling of workers in the automotive sector."

Partner with nonprofits to reach diverse talent

Many reskilling nonprofits work with populations that are underrepresented in the workforce. By teaming up with these nonprofits, companies can significantly expand access to talent and employment opportunities in ways that benefit both parties, often at low cost. Some of the ongoing reskilling efforts we learned of in our research involve corporate partnerships with such innovative entities as OneTen (which helps Black workers in the United States), Year Up (which helps disadvantaged youths in the United States), Joblinge (which helps disadvantaged youths in Germany), and RISE 2.0 (a BCG program that helps workers in Singapore without a digital background move into digital roles). Year Up stands out among these initiatives for its careful use of statistical techniques to study the impact of its training on participants. Since 2011 the program has placed more than 40,000 young people in corporate roles and internships that would have been inaccessible to them without the reskilling support and network it provided. The program has an 80% placement rate at more than 250 participating companies.

Partner with local colleges and training providers

Companies have a lot to gain by teaming up with educational institutions in their reskilling efforts. Examples of such partnerships include the UK-government-funded Institutes of Technology, which bring together colleges and major employers to provide practical technical training for workers without tech backgrounds, in ways that allow companies to quickly react to new technologies and meet rapidly evolving skills needs; and BMW's collaboration with the German Federal Employment Agency and the Association of German Chambers of Industry and Commerce, which supports the transition to electric vehicles with reskilling programs aimed at industrial electricians.

· · ·

Many companies have an intuitive understanding of the need to embrace the reskilling paradigm shifts discussed in this article, and some, admirably, have already made tremendous commitments to doing so. But their efforts are hampered by two important limitations: a lack of rigor when it comes to the measurement and evaluation of what actually works, and a lack of information about how to generalize and scale up the demonstrably successful features of reskilling programs. To adapt in the years ahead to the rapidly accelerating pace of technological change, companies will have to develop ways to learn—in a systematic, rigorous, experimental, and long-term way—from the many reskilling investments that are being made today. Only then will the reskilling revolution really take off.

Originally published in September–October 2023. Reprint R2305C

Discussion Guide

Are you feeling inspired by what you've read in this collection? Do you want to share the ideas in the articles or explore the insights you've gleaned with others? This discussion guide offers an opportunity to dig a little deeper, with questions to prompt personal reflection and to start conversations with your team.

You don't need to have read the book from beginning to end to use this guide. Choose the questions that apply to the articles you have read or that you feel might spark the liveliest discussion.

Reflect on key takeaways from your reading to help you adopt the ideas and techniques you want to integrate into your work as a leader. What tools can you share with your team to help everyone be their best? Becoming the leader you want to be starts with a detailed plan—and a commitment to carrying it out.

1. Chapter 1 warns that "big bet" digital initiatives often fail badly, whereas smaller, more incremental moves have a better chance of success. Has that proven true at the organizations you've worked at? What have you observed about the factors that help a digital initiative succeed or fail?

2. Think about a recent digital initiative on your team. Did it merely enhance an existing product or service, or did it explore new methods of creating value through data or platforms? What mindset shifts could your team make to improve future digital efforts?

3. Is digital expertise in your organization confined to IT and other technical teams, or more evenly spread across functions? How does this help—or hurt—innovation? What could leadership do to help all employees build their digital capabilities?

4. As AI gives people easy access to expertise, what unique capabilities or knowledge does your team still possess? How could that change your problem-solving or decision-making? What are you doing to protect or evolve that edge?

5. One common obstacle to digital transformation is the difficulty people have in adopting new technologies. What's one tool or system your team has struggled to use, and what made it challenging? How could your organization better help people adapt to digital innovations?

6. Chapter 4 explains that digital transformation often doesn't just mean using new technologies. Instead it involves using established technologies to meet customer needs in new ways. Have you ever seen a company introduce a digital solution that didn't align well with customer needs? What flawed assumptions were behind it? What could the company have done to improve the solution's usefulness?

7. Generative AI is enabling all employees to help redesign workflows and processes. What repetitive or time-consuming tasks could the technology help you with? What help or support would you need to make that happen?

8. What does having a digital mindset mean to you? How has your thinking about new technologies and digital

initiatives changed over time? Have any specific ideas or insights influenced your thinking?

9. How digitally savvy would you say your company's leaders are, and why? When and how has their digital expertise, or lack thereof, affected your and your team's work?

10. Does your organization give employees full, robust access to data—such as 360-degree views of customer information or in-depth views of a sales channel—or is your data access fragmented and siloed? What kinds of data that you currently lack access to would help you do your job, and what opportunities for innovation could it unlock?

11. How confident are you in working with data? What kinds of learning opportunities (formal or informal) could help you increase your confidence? Are there any specific aspects of working with data that you'd like to learn more about?

12. Do your company's interactions with customers tend to be occasional or continuous, and why? How could having digitally enabled, more-continuous customer relationships improve your products or services? What would need to happen to enable those kinds of connections?

13. Reflect on a time when your team tested a new idea. Was the process structured like an experiment (with a hypothesis, control, and measurement), or more informally? How might that structure have affected the results? What could your team do to test ideas and run experiments more effectively—or more often?

14. How effectively does your organization support reskilling efforts for employees? Is reskilling visible and encouraged?

Which groups or teams provide support, and who else could provide support that you would find useful?

15. What other sources on leading digital transformation have had a significant impact on your work? Were there voices or subtopics you missed in this collection? Were there voices or subtopics included that surprised you?

16. After reading and reflecting on this book and discussing it with people on your team, write down the ideas and techniques you want to try. Think of how you might experiment and implement them in both the short term and long term. Draft a plan to move forward.

About the Contributors

Iavor Bojinov is an associate professor of business administration at Harvard Business School. He is also a faculty affiliate of Harvard's statistics department and the Harvard Data Science Initiative.

J. Yo-Jud Cheng is an assistant professor of business administration in the Strategy, Ethics, and Entrepreneurship area at the University of Virginia Darden School of Business.

John Corwin is the general manager for corporate strategy and development at Microsoft.

Paul R. Daugherty is the AI advisory chair at TPG and the former chief technology officer of Accenture. He is a coauthor (with H. James Wilson) of *Human + Machine, Updated and Expanded* (Harvard Business Review Press, 2024).

Veeral Desai is a senior adviser to QuantumBlack, AI by McKinsey & Company.

Leila Doumi is a PhD candidate in the Strategy unit at Harvard Business School.

Tim Fountaine is a senior partner in McKinsey's Sydney office.

Cassandra Frangos is an expert in CEO succession and executive development at Spencer Stuart, partnering with boards and CEOs to assess and develop potential successors and unleash top team performance.

Nathan Furr is a professor of strategy at INSEAD and a coauthor of five bestselling books, including *The Upside of Uncertainty*, *The Innovator's Method*, *Leading Transformation*, *Innovation Capital*, and *Nail It Then Scale It*.

Sagar Goel is a managing director and partner at Boston Consulting Group, Singapore, and a global insights leader at the BCG Henderson Institute.

Boris Groysberg is a professor of business administration in the Organizational Behavior unit at Harvard Business School and a faculty affiliate at the school's Race, Gender, and Equity Initiative. He is a coauthor (with Colleen Ammerman) of *Glass Half-Broken* (Harvard Business Review Press, 2021).

David Holtz is an assistant professor in the Decisions, Risk, and Operations division at Columbia Business School and a research affiliate at the MIT Initiative on the Digital Economy.

Marco Iansiti is the David Sarnoff Professor of Business Administration at Harvard Business School, where he heads the Technology and Operations Management unit. He has advised many companies in the technology sector, including Microsoft, Facebook, and Amazon. He is a coauthor (with Karim Lakhani) of *Competing in the Age of AI* (Harvard Business Review Press, 2020).

Ramesh Johari is a professor of management science and engineering at Stanford University and an associate director at Stanford Data Science.

Orsolya Kovács-Ondrejkovic is a partner and associate director at Boston Consulting Group, Zurich, and a former ambassador at the BCG Henderson Institute.

Karim R. Lakhani is the Dorothy and Michael Hintze Professor of Business Administration at Harvard Business School and the chair and a cofounder of the Digital Data Design Institute at Harvard University. He is a coauthor (with Marco Iansiti) of *Competing in the Age of AI* (Harvard Business Review Press, 2020).

Jens Lauterbach is a professor of information systems in the Faculty of Computer Science at Augsburg University of Applied Sciences. His focus areas are the digital transformation of organizations in general and the implementation and use of enterprise technologies in particular. He also works as an independent adviser for digital transformation projects.

Hippolyte Lefebvre is an assistant professor in management information systems at University College Dublin. He studies how better data management helps organizations create value.

Christine Legner is a professor of information systems in the Faculty of Business and Economics, University of Lausanne. She is the cofounder and academic director of the Competence Center for Corporate Data Quality, where she and her research team collaborate directly with industry experts to develop concepts, tools, and methods that advance the field and practices of data management.

Paul Leonardi is the Duca Family Professor of Technology Management at the University of California, Santa Barbara, and advises companies about how to use social network data and new technologies to improve performance and employee well-being. He is a coauthor (with Tsedal Neeley) of *The Digital Mindset* (Harvard Business Review Press, 2022).

Yang Li is the director of corporate strategy at Microsoft.

Rita McGrath is a professor at Columbia Business School and a globally recognized expert on strategy in uncertain and volatile environments. She is the author of *The End of Competitive Advantage* (Harvard Business Review Press, 2013) and *Seeing Around Corners*.

Ryan McManus is the CEO of Techtonic.io and a globally recognized expert on digital business models, transformation, and ecosystems. He is on multiple public and private boards and is a contributing lecturer at Columbia Business School.

Benjamin Mueller is a professor of digital business at the University of Bremen. He specializes in corporate digital responsibility, as well as in how advanced information and communication technologies transform organizations and can augment individuals' work in innovative ways.

Satya Nadella is the chair and CEO of Microsoft.

Tsedal Neeley is the Naylor Fitzhugh Professor of Business Administration and the senior associate dean and chair of the

MBA program at Harvard Business School. She is a coauthor (with Paul Leonardi) of *The Digital Mindset* (Harvard Business Review Press, 2022) and the author of *Remote Work Revolution*.

Kayvaun Rowshankish is a senior partner in McKinsey's New York office.

Raffaella Sadun is the Charles E. Wilson Professor of Business Administration at Harvard Business School and a cochair of its Managing the Future of Work project.

Sven Schmit is a member of the technical staff at OpenAI. Previously he was the head of statistics engineering at Eppo, an experimentation platform vendor.

Andrew Shipilov is a John H. Loudon Chaired Professor of International Management at INSEAD. He is a coauthor of *Network Advantage*.

Nicolaj Siggelkow is a professor of management and strategy at Wharton and a codirector of the Mack Institute for Innovation Management. He is a coauthor (with Christian Terwiesch) of *Connected Strategy* (Harvard Business Review Press, 2019).

Mohan Subramaniam is a professor of strategy and digital transformation at IMD Business School. He is the author of *The Future of Competitive Strategy*.

Jorge Tamayo is an assistant professor in the Strategy unit at Harvard Business School.

Elizabeth A. Teracino is a senior researcher and adviser in data, digital, and sustainability transformations of the Faculty of Business and Economics (HEC) at the University of Lausanne and at the Competence Center for Corporate Data Quality.

Christian Terwiesch is a professor of operations and innovation at Wharton and a codirector of the Mack Institute for Innovation Management. He is a coauthor (with Nicolaj Siggelkow) of *Connected Strategy* (Harvard Business Review Press, 2019).

Martin Tingley is the head of experimentation for Windows at Microsoft.

H. James Wilson is the global managing director of technology research and thought leadership at Accenture Research. He is a coauthor (with Paul R. Daugherty) of *Human + Machine, Updated and Expanded* (Harvard Business Review Press, 2024).

Bobby Yerramilli-Rao is the chief strategy officer at Microsoft.

Index